Worlds of Desire, Realms of Power

A cultural geography

Pamela Shurmer-Smith
and
Kevin Hannam

Edward Arnold
A member of the Hodder Headline Group
LONDON NEW YORK MELBOURNE AUCKLAND

To John

. . . we always make love with worlds.
(Deleuze and Guattari, 1983, p. 294)

First published in Great Britain 1994 by
Edward Arnold, a division of Hodder Headline PLC,
338 Euston Road, London NW1 3BH

Distributed in the USA by
Routledge, Chapman and Hall, Inc.
29 West 35th Street, New York, NY 10001

© 1994 Pamela Shurmer-Smith and Kevin Hannam

All rights reserved. No part of this publication may be reproduced or transmitted in any form or by any means, electronically or mechanically, including photocopying, recording or any information storage or retrieval system, without either prior permission in writing from the publisher or a licence permitting restricted copying. In the United Kingdom such licences are issued by the Copyright Licensing Agency: 90 Tottenham Court Road, London W1P 9HE.

British Library Cataloguing in Publication Data
A catalogue record for this book is available from the British Library

Library of Congress Cataloging-in-Publication Data
A catalog record for this book is available from the Library of Congress

ISBN 0 340 59217 6

5 4 3 2 1 98 97 96 95 94

Typeset in 10/12 Century Schoolbook by
Fakenham Photosetting Ltd, Fakenham, Norfolk
Printed and bound in Great Britain by
J W Arrowsmith Ltd, Bristol

Contents

Acknowledgements

In writing this book we have constantly drawn upon the stimulus and support provided by groups of third-year cultural and postmodern geography students at the University of Portsmouth. They acted as a buffer against our more practical colleagues, though they didn't always give us an easy ride.

The Social and Cultural Geography Study Group of the Institute of British Geographers has been a group we felt we *belonged* to, rather than were just members of. In particular Phil Crang and Chris Philo have always been encouraging about what we are trying to do in this book. Pam has many old debts to former teachers who complained about her reticence about going public. Mary Douglas has remained a constant inspiration and probably does not realize the depth of my admiration for her. José Cutilleiro, Julian Pitt-Rivers and Jean La Fontaine all imposed their very different marks on my thinking and the late Jaap van Velsen was like a godfather. At Portsmouth, Kelvyn Jones has been the most generous and gently supportive colleague anyone could imagine. Katy Bennett has explored ideas about transgression with us and has been a practical help too. Louis Shurmer-Smith occupies a special place in the scheme of things. As a colleague, he has been stimulating; as a teacher to Kevin, he fostered an interest in the spectacular landscapes of Paris; as Pam's husband, he has made space, put up with a near total breakdown in household order, encouraged, listened to interminable monologues, argued and read several versions of what is written here. It is impossible to thank him enough.

For permission to use the poem *Cultural Studies* by Simon Armitage, we thank Faber and Faber. We are grateful to Yinka Gbotosho for permission to reproduce Fig. 7.1 to Jason Murrin for Fig. 13.3; and to the London Transport Museum (with whom the copyright rests) for Fig. 9.2.

At Edward Arnold, Laura McKelvie has been cajoling, patient and reassuring and we would never have started without her.

The book is dedicated to John Shurmer-Smith who has been good-natured about his mother's obsession with 'The Book' and who has helped with information about computer games and a young person's view of the world.

List of Figures

List of Boxes

Introduction

In this introduction we aim not only to introduce our book and our intellectual position but also ourselves, because it is an important assumption underlying everything that we write that the personal and the subjective cannot be held constant when thinking and communicating. We believe that because there is no possibility of objectivity, particularly when addressing cultural matters, it is essential that readers know *who* is writing, whose subjectivity they must filter the reading through. However introducing oneself is no simple matter: Deleuze and Guattari begin their seminal work *A Thousand Plateaus* (1988) with the words, 'The two of us wrote *Anti-Oedipus* together. Since each of us was several, there was already quite a crowd'. As you read this book you will realize just how much both of us admire Deleuze and Guattari, their rejection of simple structures, their questioning of rationality, their prioritization of desire in the interpretation of the world. We also admire their corporateness – the way in which they appear as a single author made up of lots of selves and we aim in this book to emulate this. We hope that it is difficult to unravel the Shurmer-Smith from the Hannam and that our writing is truly dialogical, a continuous thread, not a structured garment.

We are, however, two people occupying the same and different worlds. Pam Shurmer-Smith is a lecturer in geography and anthropology at the University of Portsmouth where Kevin Hannam is a part-time research assistant. There is a huge difference in our ages, incomes, family roles, working lives, gender, backgrounds and yet there is also an incredible sameness in our intellectual positions, attitudes and tastes. We spend a lot of time together, not just when we are working. We want to emphasize this sameness and difference because it is a cardinal belief of ours not only that people can generate joint ideas, but also that the easy explanation of behaviour and thought by recourse to categories based on economic and social structure is something which needs to be rejected. Throughout this book we are interested in difference, how people create it and use it to structure their systems of power and to give way to or tame their desires. We maintain that it is in the act of differentiating that repression begins, for 'appropriate' behaviours, thoughts, degrees of autonomy are granted according to constructed categories of age, gender, race, ethnicity, class and sexuality; along with notions of the appropriate go notions of both appropriation and inappropriateness. This means that we are also interested in the ways in which the boundaries between these categories are configured in space, and the ways in which they are transgressed or subverted.

This concern is not just a personal one. We are not just trying to explain that a middle-aged, married woman with a school-aged child and a career, who is a home-owner and car driver, can have a similar perspective on the world as a young, single man, who lives in a rented room and doesn't drive; we are claiming that, as we move out of the modernist era, the thinking which generates the differentiating categories we have just used is becoming increasingly unhelpful. We are saying that all those textbooks which do not problematize social categories are not just boring, they are also misleading, because they give simple answers to difficult questions: they strip the world of its enchantment but, worst of all, they perpetuate repressive thought. We obviously acknowledge that there are differences of experience and of opportunity, what we are rejecting is the morality in constructing types of people out of them. Making this move out of structured thinking is more difficult than it might seem since it holds up to question many of the 'of course' assumptions we use in our everyday lives. But these are the assumptions which we are often made to feel ashamed of if we blithely employ them in the presence of people who are hurt and resentful when confronted with the whole set of prejudices which are commonly attributed to one of their characteristics.

We are acutely aware that the world we are living in is changing very fast but our understanding of this change is not that of simple transformation, where one structure is assumed to give way to the succeeding one on the basis of a shift in the mode of production. Instead, we subscribe to the view that the change is so pervasive that the very idea of rigid structure is, in many contexts, questionable; that a state of flux is now normal and is not to be regarded simply as the unstable point of transition from one stable state to another. This means that our account of cultural geography will be from the perspective of poststructuralist philosophy, not because it is fashionable, but because we find that it really does help us to understand the world. For us the word 'becoming' is more important than 'being', as everything seems negotiable, contingent, incomplete. The ideas which were developed to explain supposedly static conditions become worthless when the relationships, values, behaviours we are studying steadfastly refuse to remain within established boundaries.

We are not at all sure that the neat structures which are associated with modernist thought *ever* really reflected the world as it was. What they did reflect was the way modernist thinkers found most congenial to construct it. By no means all people in the modernist era were (are?) modernists in their world view, but for far too long other ways of thinking have been seen in academic circles as irrational, in a perjorative sense. Here we are already becoming political, for we are recognizing that ways of categorizing, structuring and thinking are generated sympathetically to those whose greatest comfort is supported by them – comfort, not only in the sense of material conditions, but also in intellectual and moral states. The world is a more comfortable place when the legitimized view of it coincides with one's own interests, when one perceives oneself to be at the centre and others at the margins, when one's own notions of hierarchy, morality, order and intelligence do not have to be strenuously defended at every turn. The problem is that only a tiny minority of people find themselves in this enviable situation. Many women, black people, young people, old people, poor people, disabled people, people who do not belong to a 'normal' family and people whose sexuality is non-conformist find themselves excluded from the active structuring process, confined to a view of the world which always decentres them. However, though we are political in our thinking we are not partisan or advocatory; we have no cause other than that of deconstruction; there is little liberal 'niceness' about either of us and to paraphrase Nietzsche (1980), our cynicism is very close to honesty (see Box A).

BOX A Deconstruction

Deconstruction is neither analysis nor critique. It should be viewed as a strategy against the sort of thinking that asks 'what' instead of 'how' (Derrida, 1981, 1991; Bennington, 1989). As a term, however, deconstruction is fast approaching the kind of controversial popularity that has led other buzz words associated with postmodernism (see Box K) into the realm of meaningless catch-alls.

In architecture, deconstruction has been applied in the redevelopment of the *Parc de La Villette* to produce a textual language based on ideas of spatial turbulence. In fashion, designers have renamed themselves as 'architects', and 'deconstruction' means the pulling apart of clothes, chopping up three old dresses and sewing them back together into one so that the fabrics and seams mismatch. In pop music, deconstruction has become associated with a destabilizing of conventions and the highlighting of performance rather than representation. For example, Bloomfield (1993, p.25) argues that the rock group *My Bloody Valentine* is a prime example of deconstruction:

> their guitar sound is enveloped in what is at times a haze, or then a roar, or a shrieking wind of feedback and distortion ... these are guitars that are deconstructed. The sound is taken apart and reassembled against the grain of the instrument. As the band put it themselves, they aim to make their guitars 'sound as though they are not really there.'

The metaphor is appropriate. The French philosopher Derrida (1981)

Cont.

Box A cont.

has been the key proponent of deconstruction and has attempted to theorize this tension of not being there, of the absence, wandering and desolation to be found in any cultural text, whether it is a landscape, image or simply a word. From Derrida, deconstruction has been translated by Norris (1990) and Eagleton (1986) as a strategic reversal and reinscription of the hierarchical oppositions to be found in texts. This form of deconstruction was taken up by the geographer Harley (1991) in his work on maps, but as Doel (1993) warns us, deconstruction is much more spatial than it has been treated so far, it is about the impossibility of finding a final position within the text, in short, about the intervention of a strategic borderlessness in geography.

So far we have only used geographical terms metaphorically – 'world view', 'centre', 'margin', 'boundary' – and throughout this book we will emphasize the interaction between metaphor and materiality, showing not only that the way in which people think the world has very real repercussions for the way it is, but also that the way in which they experience their environment has implications for how they construct metaphors. We do not believe that culture and communication are mere epiphenomena of material reality but neither do we believe that culture is a superorganic thing which writes upon landscapes (see Box B).

BOX B The superorganic view of culture

The superorganic view of culture was never appropriated by British anthropology and was one of the main reasons why American cultural anthropology and British social anthropology developed as two such very different subjects. Kroeber in *The Nature of Culture* (1952) outlined a theory of culture which assumed that cultures were things in themselves, rather than products of individual human interactions. We can regard this view as a reification of culture. He saw culture as given, in the same way as a physical environment is given, and assumed that where change took place it was as a consequence of a diffusion of cultural traits alongside a movement of people. Kroeber believed that in explaining human behaviour it was important to isolate different separate levels of causality (1) the inorganic (geology, geomorphology, climate, etc.); (2) the organic (flora and fauna); (3) biopsychic (human

Cont.

Box B cont.
individual psychology); and (4) the superorganic culture (learned behaviour). These levels he saw as distinct, but in communication with one another, making the behaviour of people in different places.

This was a view of culture which was palatable in an American context which preferred to see the different cultures of indigenous Americans, former slaves and differently originating descendants of immigrants in terms of their own superorganic cultures, rather than contemplating the possibility that the particular interaction of people with different life chances would itself have an enormous effect upon cultural practice. The superorganic view dismissed the interpretation of differentiation emerging from internal conflict and was an inherently apolitical stance (Duncan, 1980).

Kroeber's ideas were influential in American cultural geography, being incorporated into the subject by Sauer (1963) who famously maintained that 'Culture is the agent, the natural area is the medium, the cultural landscape is the result'. Sauer further said in his Association of American Geographers presidential address that the task of the cultural geographer then became the isolation of the superorganic cultures which had resulted in particular landscapes. This is a view of cultural geography which still continues to inform popular and elementary representations of human variation, but the main objections to the superorganic view are:

(1) that it sees culture as a given, rather than a process;
(2) that it sees culture as externally causal or obstructive in terms of personality, art, economics or politics;
(3) that it postulates a spurious holism of culture which transcends the actions, interactions, interests, passions and interpretations of individuals, who are seen as the creations and carriers of their culture, not the creators of it; and
(4) that it leads cultural geography in the direction of description rather than interpretation.

The superorganic view meant that cultural geography was peripheralized in British geography, only gaining a respected position once it became possible to theorize culture in terms of human interaction and communication.

We do, however, maintain that culture is very important. It is that negotiated intersubjectivity which allows human beings as individuals to reach a tenuous understanding of one another, to experience each other jointly, to fuse the molar with the molecular (see Box C). We see culture as endlessly generating and breaking down groups of people who see themselves as more or less temporarily 'the same'. Culture for us emerges from human 'needs', 'wants' and 'desires', but we do not regard these words as synonyms for one another, though they are very close and there is some slippage in their use and meaning. It is in distinguishing between these three words that we feel that it becomes possible to generalize about radically different cultures and the way in which they are constructed.

BOX C Molecular and molar

Deleuze and Guattari can help us to theorize, in a non-hierarchical way, how the individual is composed within categories, through their notions of the molar and the molecular (Massumi, 1992). These should not be thought of as a duality, however, but as stitched distinctions, as a series of chaotic, invested and overcoded flows: 'there is not one molecular formation that is not by itself an investment of a molar formation' (Deleuze and Guattari, 1983, p.340). The molar mass is the statistical level, the level at which surveys are done by most geographers, but it is also the organic level of gregariousness, the level at which people interact and play with one another. The molecular level is best thought of as the product of processes of individualization. In contemporary societies it demassifies as processes of exploitation, control and surveillance become more subtle and diffused (Deleuze and Guattari, 1980). Privatization and personal privacy in the place of corporatism and community may be seen as molecularizations. Flows that were formerly out in the open now take place behind closed doors.

Although needs, wants and desires can be found in all human beings, we find varying degrees of significance attached to each in different settings. Needs are for those material and immaterial things without which, in the most extreme cases, life itself cannot be supported and, in less extreme situations, social life is impossible. The first half of this definition is easy enough to understand, as it relates to absolute requirement for food, shelter, human support and the like, the second half is more problematic since it must take account of the element of selectivity between social units, so Cinderella *needed* a gown to go to the ball, she *needed* to go to the ball in order to marry the prince, but it is arguable whether she *needed* to marry the prince at all (though she perhaps *needed* to marry someone in order to be a fully accepted adult woman in the context of her society). At this point we confront wants. Wants are based upon optimizing one's satisfactions, trying to get the most out of life, though these can be altruistic, like wanting the greatest happiness or success for one's children or wanting equal opportunities for all people. Wants are generated when there is choice within a situation and it is out of the notion of wants and the maximization of satisfaction that the whole of both Marxist and neoclassical economic theory is constructed. But desires are very different things, they emerge from the self, they are sensory, psychic, sexual, but they attach themselves to persons and things. We can assume that Cinderella *needed* to marry, *wanted* to marry well, *desired* the particular man. It is questionable whether one can ever maximize the satisfaction of desires since the self is never a completed project. Desires are inherently selfish and self-indulgent. Whether one considers this to be a good or a bad thing rests entirely upon one's ethical position. There are those who believe that giving way to one's desires is immoral, especially when the needs of others have not been satisfied; there are others who believe that not to attend to one's desires is to repress what is most human in us; and there is that great majority who oscillate between the two extremes of puritanism and excess, frustration and guilt with regard to desire (see Box D).

BOX D Desire

Freud mapped desire 'through the classical projector on to a single screen, the black and white Oedipal screen'. But, 'what characterizes desire is that it is a fluid system constantly inter-connecting with other surfaces and economies' (Campbell, 1992, p.97).

Following Jacques Lacan (1977), we can see that desires tend to be supported by our biological needs for certain things, whilst our wants are generated through our language, as responses to an appeal to a particular need. It is the gap between our wants and needs that often constitutes our desires, but having said that, desire is not merely an appetite, it is insatiable. Deleuze and Guattari (1983, pp.25–6) depart from Lacan's tripartite conceptualization around the lack of an object (namely the phallus) to theorize desire as a much wider force in society:

> To a certain degree, the traditional logic of desire is all wrong from the outset: .... From the moment that we place desire on the inside of acquisition, we make desire an idealistic (dialectical, nihilistic) conception, which causes us to look upon it as primarily a lack: a lack of an object, a lack of the real object. ... Desire does not lack

Cont.

Box D cont.

> anything; it does not lack its object. It is, rather, the subject that is missing in desire, or desire that lacks a fixed subject; there is no fixed subject unless there is repression.

Deleuze and Guattari are simply arguing that there are many subjective positions of desire and what we desire at the molecular level is not always what we desire at the molar. This may seem like a contradiction, but as they advise us, no one has ever died of contradictions or even mild transgressions come to that.

We do not want to be too mechanistic in this, but there seems to be some degree of 'fit' between the *prioritization* of needs, wants or desires in particular societies and their economic and political structure. Generally speaking, we can assume that the orientation towards needs will be greatest in societies with subsistence economies, where households produce most of their requirements for themselves and there is not a great deal of variety in goods or ways of living – the societies which have often been referred to as tribal and peasant. We associate a preoccupation with wants with modern industrial society, where production is directed towards the market and exchange; indeed the whole idea of maximization of satisfactions turns upon the ability to evaluate different things and experiences in terms of one another, usually through the medium of money. Desire comes into prominence when a society moves into a post-industrial stage, when the communicative and expressive value of goods rises above their use value, when experience is more highly rated than consumption and where the cultivation of taste, preference, ambiance is aspired to. It seems foolish to us to try to understand economies of desire with a logic which was generated to explain economies of want, for where want is rational, desire is irrational; need is quite outside the rational/irrational divide and always *seems* to belong in the realm of nature.

Here we come close to the work of the French Marxist anthropologist Godelier (1986), who demonstrated that the place of the economy shifted in different societies. Godelier had difficulty in reconciling his Marxist ideology with his ethnographic knowledge of differently structured societies and, in working through this intellectual problem, he threw considerable light on the way in which value and values are constituted. He saw that the way in which production and consumption were controlled would influence the set of values which would be seen as central in society – so where a group of related people produced for themselves, there would probably be a high esteem for kinship-based values; where a priestly group controlled production, there would be an emphasis upon religious values; where military might translated itself into wealth, political power would seem to explain the ideology; only where market forces dominated production would there be an overt centring of the economy as a thing in itself.

We are labouring this point because we feel that much geography until recently has unquestioningly applied the logic of the industrial phase in situations where it was not necessarily appropriate; that geography has been narrowly economically driven in its explanations of human behaviour and its spatial outcomes. By this we mean not only that geography has prioritized considerations of production, distribution and consumption in market conditions in its constructions, but also that the rationality which emerges from an economics of maximization has influenced the way in which geographers interpret decisions about the use of space and relationships with the environment. We are saying that the characteristics of the economy will influence the value systems of a society, but that it is quite possible for these value systems themselves to deny the centrality of economistic thinking. So we find in postindustrial societies that significant numbers of people believe that leaving the environment at the mercy of market forces is not only morally wrong but also dangerous; the postmodern attitude to the environment contains a strand which assumes that the utilitarian evaluation of resources should take a second place to unquantifiable ideas about well-being, intrinsic worth and sentiment. Although it is sometimes expedient for bargaining purposes to try to put a monetary price on these, to do so is not only to do violence to their value in a completely different realm but also to acknowledge the superior force of market-oriented thinking. In our economics we insist that there is room for the terms 'priceless' and 'invaluable', and we do not believe that these mean the same as 'of very great money value'. Whilst locational analysis and the like has been clearly influenced by market forces, we can see in environmentalism an attempt to break free from the systems of market value and into a new ethics, but this is often marked by a very real difficulty in finding a persuasive logic and language. Rational thought based upon the economy of want is so pervasive in academic and political circles that other ways of evaluating are often bracketed as marginal, woolly, unrealistic or trivial. If one considers the way in which the word 'sentimental' is used as a term of abuse, the power of the economy of want becomes instantly apparent; we are saying that values constructed around sentiment are as interesting when making interpretations of geographical phenomena as values based around wants.

In the past cultural geography has had a rather ambivalent status; often seen as secondary to a geography predicated on locational or spatial science, there is no avoiding the fact that at times is has been regarded as a low-prestige, hobbyist activity for the self-indulgent or the amateur geographer. These days are irrevocably gone and it is arguable that in the future an unsophisticated stance on culture will be as much of a disability for a geographer, of whatever subdisciplinary persuasion, as an ignorance of economic theory has always been. We make this assertion without any degree of tentativeness because of our conviction that decision-making based upon desire rather than want will increase, and not only in relatively affluent Western societies. The numbers of people locked into need will probably increase too and, perhaps paradoxically, for those who find their needs difficult to satisfy, desire, mediated through advertising, television or spectacle, may seem more seductive than want. We will explore this theme more fully in later chapters.

We see the contemporary cultural geography (we are hesitant about using the term 'new' because there are continuities with the past) as being concerned less with the mapping of cultural traits onto landscapes than with expressing, at the molecular and molar levels, ways of being in the environment. We use the word 'environment' in its widest possible sense as the world in which one lives, with all of the multiple meanings of the verb 'to live'.

We concentrate rather a lot on novels, films, poems, contemporary philosophy, fashion and lifestyle because we are certain that it is through these that we can appreciate the multitude of different ways in which people conceive of themselves in the world. This being in the world is not necessarily a comfortable experience, it is not invariably about being attuned to one's environment but can just as easily be about feelings of alienation, marginality, isolation, claustrophobia, entrapment, rejection, anger, boredom, resentment or fear. These feelings about being in the world have further implications for mundane decision-making about choice of residence, micro-location of industry, tourism and retailing than do simple notions of monetary cost – indeed they can translate themselves into prices if widely experienced. Though only a minority of people express themselves in the creation of communicative works, everyone expresses and constructs the Self in the consumption or contemplation of them.

A novel or a film can catch the way people feel, but it can also shape the constitution of thinking, feeling and evaluation and it can influence the way that people act. A creative work is not a *mere* representation, though representation is a part of what it is; it is also imaginative, though the imagination will always be, more or less, within the bounds of possibility of the culture which nurtured it. Whilst such works enable us to think about our own and other times and places, we need to be very cautious about the way in which we use them; those who have regarded them as documents of 'reality' miss the fantastical element in any communication; those who see them as nothing more than fictions can miss the point that fiction, like dream, is created out of experience and longing. We are less sanguine than many about the relationship between reality and fantasy and would be reluctant to ascribe the direction of causality between the two or even to try to tease them apart.

It is for this reason that we have put aside those vulgar Marxist views of culture which rely upon the notion of culture as a superstructure resting upon a base which consists of the mode of production and its corresponding relations of production and political order. For Marx a culture, particularly in its religious aspects, represented the world view of the dominant class and when it was accepted by subordinate classes could only be understood as false consciousness. Culture would only escape from this when, with the abolition of classes through class struggle, there would only be culture based upon objective materiality, practice and representation of practice. (Hence the rise of both constructivism and socialist realist art and the moral elevation of rationality above religion in the former Soviet Union.) We do, however, have respect for the position of Gramsci (1971), the Marxist cultural theorist who introduced the notion of cultural hegemony, the process whereby those forces which have power within society will gain control of the culture, in part by means of the media and education. For Gramsci hegemony implied within it some degree of consent on the part of the majority of society and the construction of a 'common sense' which tempered the constructions of those in power. He assumed that cultural hegemony had to take some account of the aspirations and preferences of those who were not powerful, otherwise, the communication would be bracketed as mere propaganda, and rejected, rather than as culture, and attended to. As we stated earlier in this introduction, we also find Godelier's (1986) Marxism useful; he makes us rethink the idea of superstructure, insisting that, in Marx's native German, *Überbau*, the word invariably translated into English as 'superstructure', means a construction, an edifice which rises up on foundations, the *Grundlage*, we then live *in* the *Überbau*. This is an interesting shift from the assumption that we can only look up to a superstructure, which becomes, by implication, superfluous to the mundane. So for Godelier culture is not the epiphenomenal frills upon structure, it is part of structure, the whole resting upon a foundation which is the mode of production; it is certainly not a simple confidence trick played by those with power upon those without it, though, clearly, power relations will inform culture.

These ideas of Gramsci and Godelier are in the back of our minds as we think through our cultural geography, but we find their grand theoretical positions rather blunt instruments when we look at the local experiences of people. Grand theory has it uses at the very large scale but its explanatory power diminishes when one comes down to the particular, at which point it becomes necessary to pay attention to the genealogical, the biographical, the somatic and the psychoanalytical. In the contemporary world there is an increasing awareness that generalization across broad swathes of humanity has little relevance to life as it is lived. Just as retailers have realized that it can be more profitable to sell to small, easily targetable niches of the market than to aim for a share of the mass, so cultural geography is becoming aware that it needs to be able to understand difference rather than to construct artificial homogeneity if it is to interpret cultural matters.

In the rest of this book we will focus upon different experiences and mediations of space, place and environment, always conscious that in doing so we are leaning upon constructions which need to be problematized and that, no matter how hard we try, we will always think and write from our own vantage point, perhaps unknowingly angering others as we do so because we have omitted or misrepresented their point of view. It is our belief that, as much as possible, we should not speak for others, and so we hope that this book will not look like an attempt at a finished statement, a conclusion as to what culture *is*, but that it will help to open up space for debate and dialogue. We thus make no claim to provide an overview of current wisdom in cultural geography and certainly we would not like to think of anyone thinking of our word as 'gospel'. Culture is not a graspable thing but a continuous flow into which we can wade.

SECTION I

Beyond Sense of Place

Place is a deceptively simple concept in geographical thought; we want to make it difficult, uneasy. We want to show that places do not exist in a sense other than culturally, and as a result that they can appear and disappear, change in size and character, and even move about according to the way in which people construct them. Places then have no objective reality, only intersubjective ones (see Boxes E and F).

BOX E Xanadu

All that guff,
about place and space,
an ocean of stuff
and still it's a case

of ip dip dip,
my blue ship,
which came first

the flea or the pit?
Which makes which,
the pig or the stye?

Cont.

Box E cont.
All that time
and still we're not certain,
what wears what,
the brick or the person.

(Armitage, 1992, p.23)

At first sight this statement may seem to border on the lunatic: with some degree of certainty we tell people where we live, buy airline tickets and finish up where we expect to, choose universities partly on the basis of the congeniality of the towns they are situated in. We act as if places exist in the sense of having a fixity and an identity and we are generally not frustrated by the assumption. However, our lunacy starts to lessen when you think about all the times you have been somewhere you visited as a child and say that it is smaller, when you revisit somewhere you were in love and find that it is surprisingly mundane, when you go somewhere that you once thought was the height of sophistication and find that it is embarassingly tacky. We want you also to think about the implications for place when countries change their borders and/or their names, when streets are realigned, when buildings are knocked down, when fields are built over or when downland becomes a motorway – are they the same places with different characteristics or different places? Pam constantly confronts this problem of fluidity of place when trying to explain where she grew up – if she says 'Zambia' it gives the impression that it was in an independent African country, but Zambia was not there when she was; if she says 'Northern Rhodesia' it gives the impression that she is such a dyed-in-the-wool imperialist that she can't come to terms with the change in the name – the place becomes time-specific, not merely located. This is recognized when we talk about the site of a place, the place has gone, but you can look for the site of Troy.

Cultural, not natural, considerations create, destroy and change places, even supposedly wild places, as people in the act of naming, distinguishing and characterizing separate, say, the Alps from the Jura, the Atlantic from the Pacific. Once those wild places become associated with particular human activities, whether exploration, recreation or warfare, they more obviously become loaded with cultural meaning.

The way that people experience and conceive of places varies enormously through time, between groups and between individuals. Places can go in and out of fashion, they can go from being relatively obscure to relatively well-known, and the ideas and sentiments they evoke can vary or they may stay the same but the value placed upon them can change. So we can say that, notwithstanding their concreteness, all places are imaginary, they exist in the mind as well as on the ground. Changing speeds of communication alter not just the relationships between places but, consequently, the places themselves and the way in which they are thought as knots in networks of meanings.

In the following five chapters we will look at some of the ways in which places are imagined, how they pass into a cultural repertoire of meanings and how wider cultural values impact upon them. We will start with ideas about foreigness and exoticism; move on to concepts of 'home'; think about the ways in which some places seem to be identified with particular periods of time; look at fantasy places and finish up asking whether it is possible to think of the entire world as a single place. However, we do not see these five as separable, other than for convenience. They are not conceptually distinct, as anyone who has ever felt 'at home' in a foreign country will testify. Fantasy talks to the mundane, the past is the creation of the present and the global and the local are to be experienced whilst standing in the same place.

BOX F Place

A place (lieu) is the order (of whatever kind) in accord with which elements are distributed in relationships of coexistence. It thus excludes the possibility of two things being in the same location (place). The law of the 'proper' rules in the place: the elements taken into consideration are beside one another, each situated in its own 'proper' location, a location it defines. A place is thus an instantaneous configuration of positions. It implies an indication of stability. (De Certeau, 1984, p.117)

Throughout these five chapters we want to emphasize that place is not a synonym for locality; that a range of people may inhabit the same locality but different places of differing size and characteristics. Here it is interesting to look at Smith's paper 'Homeless/global: scaling places' (1993), in which he asserts that 'It is geographical scale that defines the boundaries and bounds the identities around which control is exerted *and* contested' (p.101). Here what he means is that one person's home place may be a confined and beleaguered space, whereas a contiguous person, with a different set of experiences and opportunities may well feel himself (probably) to be autonomous and unbounded. Cream's (1993) paper on the way in which Cleveland has become imbued with the taint of child sexual abuse is a particularly interesting example of the symbolizing of places which are not directly experienced, but there are many other instances of places serving as metaphors for emotions, desires and sentiments.

This brings us to the potentially rather difficult concept of *Dasein*, a word usually translated as 'dwelling', but probably best left in the original German in order to mark the fact that it is used in the context of Heidegger's (1971) phenomenology. *Dasein* has connotations of oneness with place, ability to inhabit a place as opposed to simply being in it; it therefore contains ideas about appropriateness and appropriation or what Harvey (1993) summarizes as 'the capacity to achieve a spiritual unity between humans and things' (p.11). We shall explore this concept in the rest of this section, but particularly in Chapter 2 where we will contemplate the general and outsider implications of *Dasein*, particularly ideas about prospect and refuge as formulated by Appleton (1975). Massey (1993), in particular, has reacted to the notions of place based upon *Dasein*, arguing that they all too easily skate over the way in which places are processes, not single essential identities with unproblematic boundaries. It is not a very big leap to make to see that an unambiguous, bounded place would have to represent the view of a single fraction of society – *Dasein*, for all its superficial cosiness, soon slides into rather uncomfortably repressive stances.

Much traditional cultural geography has turned on the idea of 'sense of place', a belief in the uniqueness of places which characterized the idiographic strand of regional geography. This idea is very close to that of *Dasein* and contains within it the assumption that 'authentic' places are those which human activities are attuned to their environment. We can see sense of place emerging from the Vidalian regional geography based on the idea that *pays* with distinctive material cultures and traditions could be isolated on the basis of interaction of people and their physical landscape; it is dependent upon a view of rural society as relatively unchanging and regions as more or less bounded. Sense-of-place studies (particularly good examples are Relph, 1976, and Lynch, 1972) tend to be anti-modernist, drawing upon conventionally traditionalist views which hark back to Tönnies (1955) ideal type of *Gemeinschaft* or community. This section is called 'Beyond Sense of Place' because we want to push beyond the idea that particular places have their own spirit or essence and the implication that a single *Dasein* is appropriate to particular configurations. Instead we want to explore the thought that the places that people construct are polysemous and are experienced in a multitude of ways, sometimes complimentary, sometimes conflicting, sometimes just differently.

1

Foreign places

There is no denying that many people are drawn to geography by its sheer exoticism: the impression that the whole world and its rich diversity is on offer to the geographer is a sophistry which journals like the American *National Geographic*, the French *Géo* and the English *Geographical Magazine* do nothing to dispel, with their emphasis on the spectacular in landscape and the picturesque in culture.

One line of geography's genealogy reaches back to the era of heroic exploration (Driver, 1992; Livingstone, 1993). Though this heritage is largely guarded today by the non-academic wing of the subject, and is seen in the activities of the Royal Geographical Society, there are few geographers who are totally unmoved by this aspect of their subject. The requirement in most British universities that geography undergraduates undertake fieldwork in a foreign location and the encouragement given to student expeditions can, in part, be seen as the legacy of geographical exploration and fascination for the exotic. Rose (1993a) has been particularly critical of the way in which fieldwork has been used as a device to construct the geographer as *hero* (not heroine) and Pratt (1992) shows how the act of scientific appropriation of foreign places constructs the explorer as an imperialist and part of a paternalistic system of power. She demonstrates that the apparently neutral act of imposing a Western 'order' upon chaos, by means of scientific description and classification of people and natural species, served to extend intellectual power into what was conceived of as unknown. In a review of Hanbury-Tenison's edited work *The Oxford Book of Exploration* (1993). Evans (1993, p.8), himself a travel writer, points out, on the matter of heroic exploration, that 'a hundred years ago the RGS was packed with rather more colonels than a military junta; today's personal triumph was tomorrow's imperial supply route'. It quickly becomes clear that we cannot think about concepts of foreigness without realizing that there are complicated power relations involved in the construction of the category and that to consider a place and its people as foreign is to exclude them from an assumed (Western) normality.

In the 1960s anthropology came in for a great deal of criticism (see particularly Gough, 1968 and Asad, 1973) for its status as 'the child of imperialism'. The first reactions to this accusation were of wounded resentment (after all, most individual anthropologists were benign figures who had been seen as distinctly suspect by colonial administrators) but the criticism came to be seen as more and more justified when the discipline started to consider the way in which it had unproblematically constituted the people it studied as 'Others'. 'They' became an *object* of study, rather than a speaking subject and of course their way of thinking had to be made comprehensible within the terms of 'our' logic in a process of translation and interpretation. It is when one looks back at the anthropology of former ages that one becomes aware of its project in legitimizing a dominant moral position in the world.

Stocking's *Victorian Anthropology* (1987) makes sobering reading as we see the construction of the category 'savage' serving to justify not only the imperialism which was assumed to bring about their civilization, or the slavery which could be based on an assumed inferiority, but also the continued denial of civil rights at home to those who were presumed to fall between the categories of animality and civilization, namely the labouring classes, children and women, all of whom were frequently compared with 'savages'. However, geography's own position as the military's right-hand man meant it was also bound up with the stereotyping of indigenous people as 'savages', particularly during expansionist conflicts in order to normalize behaviour that would otherwise be thought of as illegal (Miller and Savage, 1977). Military geography's masculine gaze on other cultures was often enshrined through that other instrument of power, photography. As Virilio (1989, p.7) has argued, 'the history of battle is primarily the history of radically changing fields of perception'. People in foreign countries were often assembled for their portraits to be taken for geography texts as if they were exhibits rather than human beings.

However, geography largely escaped criticism at the time when independence was granted to the majority of the countries of Britain's empire, perhaps because it was then heavily involved in demonstrating that it was a spatial science, probably because the subject did not have a tradition of self-criticism or political awareness. When one recalls the extent to which geography did *directly* serve the cause of imperialism, as well as offering the image of the intrepid explorer which must have seduced thousands of young people into the colonial enterprise, it seems ironic that the subject is only now beginning to confront, with something approaching shame, its relationship with the exotic.

It was Said, the Palestinian literary theorist, who drew the attention of the West to the insidiousness of romanticizing 'other' cultures. His book *Orientalism* (1978) has had a considerable impact well beyond its starting point in the academic study of literature. It demonstrated that the construction of a 'mystic orient' as a distinctive place, served primarily to emphasize the 'flexible, positional superiority', of the West (p.7). Said drew explicitly on Foucault's proposition that power and knowledge were intimately related through a masculine gaze on the 'Other'. But we can also see Said implicitly drawing upon Derridean deconstructionism and its notion of *différance* (this is not a misspelling!) which assumes that wherever differentiation takes place, a binary opposition is constructed, one side of which is seen as superior to the other, which defers to it. For Said, the Orient exists only in counterpoint to the Occident. It contains desires and fantasies which the West wishes to contemplate but also to think of as existing only beyond the boundaries of its own normality. The fabricated 'mystic orient' is thus often seen as a timeless place, built around notions of the eternal and the archetypal, but the timelessness is associated with the idea of the past as a tense, rather than as history as a process. It is as if we wanted to have a superficial and anachronistic view of the East, conceived of as one place in one time so that we can construct our fantasies upon it. So we find contrasting representations of the East which stress either luxury, lasciviousness, sensuousness and passion or cruelty, coldness, starkness and intrigue. It is the function of the Orient to play 'Other' to the West's 'Same'.

There is no doubting the seductive power of the symbolic system constructed by Orientalism, whether we are talking about popular cultural representations or serious academic research, but Said shows how the very notion of Orientalism as academic endeavour itself serves to devalue the societies and cultures which are seen as coming within its remit. It is only when one contemplates the impossibility of setting oneself up as an occidentalist, even outside the West, that the insult becomes apparent to many Westerners. Said recognizes that Orientalists frequently 'love' the countries they develop expertise in and yet, as privileged onlookers and commentators, they cannot escape conniving with the resulting power relations.

One does not have to look far to find evidence to fit Said's view. It is all around us in films, advertisements, magazine articles, novels, computer games and so on. It informs much of the genre of travel writing and people's choice of holidays. It insinuates itself into taste in clothes, food, interior design, perfume, and more insidiously, it also infects our political relations. Once people and their lands are constructed as outside, it becomes difficult to empathize with them; once they are located in a land of fantasy, it is a short step to seeing them as not having an equal validity.

In thinking about this, we can start with something apparently quite trivial – the best-selling computer game *Prince of Persia* – in order to understand the intertextual construction of the spectrum of ideas and attitudes that constitute Orientalism. The game is a platform/maze/beat-'em-up with sumptuous graphics of palaces, dungeons and caves. Its scenario is based around the quest of a prince to rescue a princess imprisoned by the wicked Grand Vizier who intends to marry her by force; fat eunuchs, skeletons and swordsmen in exotic costumes lie in wait for the glamorous hero – heads roll and people are impaled on spikes. Obviously all computer games oversimplify settings and stories, that is why they are interesting as an example of gross stereotyping; they get right down to the most clichéd representations, recognized by everyone. *Prince of Persia* exists in that general-purpose composite 'land' of Turkish delight, *Kismet*, *Arabian Nights*, *Aladdin*, a land of mystery and violence, sultans and slaves, genies, flying carpets, snake charmers and dancing girls (see Box G). The implicit message is that in contrast to the rational, liberal, scientific West, here is an Other which is driven by indulgence and cruelty.

BOX G *Arabian Nights*

The first verse of the song *Arabian Nights*, which accompanies the title sequence of Walt Disney's cartoon film *Aladdin* (released in Britain in December 1993), is, like the film, by any standards Orientalist:

> I come from a land, from a faraway place where the caravan camels roam.
> Where it's flat and immense and the heat is intense.
> It's barbaric, but, hey, it's home.

But they are a considerable improvement over the original American version (not withdrawn on cassette) which offended Arab pressure groups:

> I come from a land, from a faraway place where the caravan camels roam.
> Where they cut off your ear if they don't like your face.
> It's barbaric, but, hey, it's home.

In contrast to the representation of the Orient as cruel and hostile is an alternative image of it as a place of passion and sensuality. There is a group of perfumes known as Oriental, whose scents are heavy, musky, spicy and sensuous. They are blended to evoke ideas of Eastern mystery and promise and sold to the Western market with images associated with introspection, the interior and concealment. The models advertising these perfumes embody ideas about seductive rather than assertive female power and the space they occupy folds in upon itself. It is remarkably close to the princess in the dungeon whose powerlessness exerts a potentially fatal power over her rescuer. Obviously, such representations do nothing to promote any depth of understanding of power structures, sexuality or gender relations in Arab countries, they turn upon *our* fantasy, not *their* practice; but they do serve to 'feminize' the East in the sense of representing it as passive and sensory rather than active and cerebral.

The myth about Oriental mysticism and seclusion in relation to sensuality escapes beyond its initial geographical bounds and becomes a clichéd sexual motif in the West which is often incorporated back into notions of the domestic. In Chapter 7 we talk about the domestic style of the architect Adolf Loos who designed inward-looking houses which play upon fantasies of exoticism, women and concealment within a European context. But, the uses of Orientalism in Western architectural design often points towards a frivolity. One thinks of the Brighton Pavilion, with its ludicrous (literally, in the sense of playful) mélange of Eastern styles, built for the Prince Regent as a pleasure palace, but one also thinks of the large number of domed and minareted 'picture palaces' (as in Fig. 1.1) built in Britain before the last war. There is also a style of 'ethnic' interior decor, particularly popular with liberal academics and students alike, where sumptuous untidiness abounds in a setting of wall-hangings and joss sticks.

Said sees the rationale behind all these apparently benign Orientalist endeavours as being the legitimization of the imperialism of the West through a construction of a moral superiority. We can see examples of the uses to which the West puts this construction when we look at the rhetoric used in relation to Iran, Iraq, Libya, the leaders of which have often been slipped into the role of the wicked Grand Vizier. The Orientalist mode allows the construction of landscapes of fear, which then makes it possible to justify aggression as self-defence or the heroic rescue of the weak (as in the game *Prince of Persia*). Dodds (1993, pp.72–3) has argued that the 'remarkably sanitised and controlled television spectacle', of the 1991 Persian Gulf conflict was based upon the writing of a series of edited foreign-policy scripts: 'by drawing on the memories and scripts of World War II, the Western Alliance was able to depict the Iraqi invasion of Kuwait, the subsequent Allied launch of Desert Storm and the short land war that followed in highly dichotomous terms.' Just like a computer game's gross stereotypes, the geographical representations of Iraq in the conflict had to be highly arbitrary in order to succeed in justifying the aggression and loss of life. Brah (1991) has shown how Asian women immigrants were not perceived as requiring rescue in the same way as white women and we may recognize this as a contemporary form of the fear about 'white slavery'.

At the time of writing this book Islamic fundamentalist terrorist groups in Egypt are waging a campaign against Western tourists, several of whom have been murdered and many foreign offices have warned their

Fig. 1.1 An Orientally inspired cinema, converted into a nightclub

nationals of the dangers of travel to that country. Of course, Egypt has long served as a destination for those in search of the sensuous version of the Orientalist myth – Flaubert (1973) went to Egypt very specifically to experience what he saw as its decadence and, like many less articulate commentators, found exactly what he was looking for, rather than what was perhaps really there. His letters represent Egypt as if it were there entirely for his entertainment (all perfumes and courtesans) or not coming up to his standards (dreaming of water ices whilst bewailing the shortage of drinking water in the desert). He followed on from Dumas' *Tangiers to Tunis* (1959) trip which had taken place three years earlier. Dumas having written one of the most impressionable Orientalist texts ever, depicting North Africa as one sensuous delight after another, an exoticism of marksmen and horses, swords and daggers, dancing girls, chieftans, feasts, colour, silks and cosmetics – and all apparently purchasable (see also Box B). This genre of writing was picked up more recently by Durell in *The Alexandria Quartet* (1962), Manning's *Balkan Trilogy* (1981) and Lively's *Moon Tiger* (1987). The fictions of exoticism and sensuousness serve to perpetuate the image of Egypt as a locus of quasi-colonial self-indulgence. It can be understood, therefore, that those who seek to live a pure and ascetic form of Islam find it hard to come to terms with people whom they not only regard as coming from a corrupt culture, but who seem to them to be using their country as a resource for the construction of their own pleasure and patronage. Enloe (1989, p.53) makes this clear when she points out that:

> For their part, Egyptian women organizing and writing as feminists in the early twentieth century were frequently more exercised by European women's stereotypical attitudes than they were by Egyptian men's protection of male privilege. They felt compelled to defend Islam in the face of racist Orientalism. They objected to portrayals of Islamic society as incapable of dynamism and reform and to writings that pictured all Arab women as mindless members of the harem.

But as these older Orientalisms are contested, often violently, but always subtly, newer Orientalisms are emerging in the West to take their place. In seeking to come to terms with the ever-expanding financial power of Japan and the huge market and labour force in China, the United States in particular, is constructing new Orientalist discourses. If one takes seriously America's decision to turn its back upon Europe, which it is increasingly portraying as chaotic, antiquated, fragmented and self-indulgent, then East and West begin to align differently. Whilst many older people in Europe and America are taking refuge in an unpleasant revival of 'yellow peril' clichés, constructing an East Asia of callous inscrutibility, lacking in spontaneity, hard working and humourless, they seem unaware that their children and grandchildren do not just buy Japanese hardware but are re*orient*ing themselves, longing to visit the land of *Sega* and *Nintendo*, watching Japanese cartoons dubbed into

European languages, accepting as a fact of life the importance of Japan in their youth culture.

The West's attitude to Japan (and to East Asia in general) is, to say the least, ambivalent. It contains desperate gestures of the old patronizing Orientalism, which are no longer convincing given the impossibility of constructing contemporary Japan as subaltern. There is a swing between reverence and contempt, emulation and rejection, collaboration and recoil which only those who are unaware of the compromises involved in subaltern status are surprised by. The new Orientalism is constructed out of weakness and desperation and is different in motivation from the Orientalism Said documents, though it learned its attitudes in the same school. Clearly, it is a reaction to the shift of economic and technological supremacy away from the sites of modernism in the 'West'.

Japanese companies now either own outright, or have significant stakes in, most of the major 'American' film, record, and computer games companies. Sony own CBS Records and Columbia Pictures; Matsushita owns MCA Universal. There has been a $6 billion Japanese investment in that most American institution, Walt Disney Corporation. Nippon TV Network paid for $3 million of the restoration of that masterpiece of European art, the Sistine Chapel. When geographers anguish about the impact of 'Western' culture in a global context it is easy to forget that the motive power is no longer American.

Morley and Robins (1992) demonstrate in their paper on *techno-Orientalism*, that Japan, in response to the clichéd manifestations of

BOX H *Endymion*

I saw Osirian Egypt kneel adown
Before the vine wreathed crown!
I saw parched Abyssinia rouse and sing
To the silver cymbals ring!
I saw the whelming vintage hotly pierce
Old Tartary the fierce!
The kings of Ind their jewel-scepters vail,
And from their treasures scatter pearlèd hail;
Great Brahma from his mystic heaven groans,
And all his priesthood moans,
Before young Bacchus' eye-wink turning pale.
Into these regions came I

(Keats, 'Song of the Indian Maid' from *The Complete Poems*.)

Western cultural arrogance, has responded with a counterattacking mixture of racism and economic superiority. Japanese desire to acquire controlling interests in media corporations, cultural institutions and real estate rather than industry demonstrates an appreciation of the sources of power in the contemporary world. However, there is little obvious Orientalization of 'Western' culture, only a detached, globally oriented appropriation of software which is effective only in tandem with Japanese-produced hardware. Meanwhile Japan constructs the West as its decadent site of tourism.

All tourists, it seems, find a version of what they seek, apparently less aware than local people that they are, in fact, helping to both recreate and perpetuate it. What is required of Egypt by the Western tourist, and of England by the Japanese tourist is that it be foreign; foreign not in the sense of unknown, but slightly different. Ardener (1987) showed how those places which are characterized as remote invariably have a very strong image and also a considerable power of attraction. The foreignness has already become known through international media and advertising and is deliberately cherished as a commodifiable resource.

There is an argument that all tourism rests on exploitative relations, as local cultural heritage and traditional lifestyles are packaged for outsider consumption and, simultaneously, outsider lifestyles and tastes are allowed to overrun indigenous ones (Turner and Ash, 1975). The commodification of behaviour as exoticism for the paying outsider sets up a curious spiral whereby what was uniquely local is preserved for international purposes, and can lose much of its original meaning. Scraps of culture are banalized when they become performances or souvenirs. Activities which were formerly *done* are now produced only to be *watched*. They become edited, prettified and invariably torn out of context to suit international tastes. Dennis O'Rourke's movie *Cannibal Tours* features Western tourists visiting 'ex-primitives' in New Guinea whose cannibalism is the attraction. The film is a starting point of MacCannell's book *Empty Meeting Grounds* (1992), in which he shows how the timing of rituals is altered to fit in with tourist schedules, work patterns are changed to make time for performance, and spatial arrangements are manipulated so that ceremonial buildings are accessible to the tourists. Artefacts like weapons, masks and penis-sheaths are manufactured not for their original use but as curios for sale, destined to be ornaments; stalls selling them appear in formerly subsistence villages. MacCannell's message is that the tourists are the new cannibals, consuming exotic culture which is made expressly for them. The meeting between global sophisticates and assumedly 'ex-primitives' becomes one in which only difference is valued by the dominant (paying) party.

Urry demonstrates in *The Tourist Gaze* (1990) that in late capitalism the desire for experience becomes greater than that for mere enjoyment amongst the people he calls post-tourists. The foreign destination in this case becomes more valued for its 'outlandishness' than for being able to offer generally recognized pleasure and, as a consequence, more and more exoticism is offered up for consumption. In Britain 1993 was the year when Vietnam was suddenly 'discovered' as a prestigious tourist destination, replacing Bhutan which had been 1992's outlandish choice – meanwhile India had become too 'obvious' to be useful for scoring points for intrepidness! When places come to have as much meaning as 'couture' in the fashion world, a new kind of geographical imagination is at work, but the volatility of fashion can have appalling consequences for those who not only supply but *are* the product. Post-tourists not only want to see, they want to join in; but then they want to withdraw and the contact is only fateful in one direction. Part of the consequence can be seen in terms of economic impact and problems of sustainability, but there is also the whole issue of local dignity as people are taught by foreigners the meaning of foreignness and how to be 'Others'.

Accompanying the global expansion of tourism has been an increased demand for travel writing. Much of this demand is fuelled by a desire to familiarize oneself with the appropriate forms of the exotic prior to a holiday or business trip, but there is also 'armchair travelling', in a nostalgic sense for places already visited, or longingly for those which are inaccessible. Here we are not referring to guide books or accounts of places but the genre known as travel autobiography, where the experience of travelling is the substance of the narrative and the place and its people become a foil for the heroic drama. The settings differ radically and so do the adventures – some like O'Hanlon's (1984) involving threat to life, others, like Theroux (1975) offering nothing much more terrifying than bedbugs or boredom – but they all have a common theme of being an innocent abroad, surrounded by foreigners who run a gamut of stock 'types' through dangerous, importuning and amusing, to exploitative, generous and inquisitive. Travel writing exposes the author to foreignness in order to allow self-discovery, it is generally more concerned with the travelling hero than with the people encountered, who are seen through the writers' experience and are accorded little subjectivity of their own.

Travel writing, much like thriller writing (which also depends upon the notion of the alien), changes in style with political context as different aspirations for mastery configure desire. Much of the writing of the high imperial phase was preoccupied with emphasizing the gap between civilization and primitiveness. Male writers sought to impose order, if only conceptually, and whilst we can see women travellers of the nineteenth century challenging the established norms within Western societies, clearly they did little to challenge them outside their own countries (Domosh, 1992a, 1992b). If one thinks of Kingsley (1987) in West Africa, wading through swamps, sweltering in her corset – not only did she refuse to give up the respectability of the corset, she found it necessary to tell her readers about it. Or Emily Eden in 1838, fussing about getting a consignment of Paris bonnets as far as Simla. Both were magnificently resilient women, but there was no doubting that they were utterly convinced of their own cultural superiority in an alien setting (the nearest Emily Eden came to 'going native' was buying finely embroidered Kashmiri shawls, one of the first examples of what is now known as ethnic fashion).

After the 'heroic' stage of conquest, travel writing settled down into a rather laconic 'District Commissioner' mode, often wry and gently self-mocking, focusing on getting on with the job of provisioning one's bungalow and running the Empire as best one could, given the problems of understanding 'the natives' and the lack of civilized comforts. Probably the best known example of this period is Grimble's *A Pattern of Islands* (1952), a book which reinforces the normality of the colonial civil service and the *right* to be there, even though it was published five years after India gained independence. In accounts at this time the trope employed was that of the gentleman administrator, running 'his' district with a benign amateurishness. The foreign is neither glamourized nor terrifying, merely baffling or amusing, depicted as different in much the same way as the same class of man might have thought about his servants or children. It was an avuncular, tweedy, Home Counties view of imperialism which sold best just after the Second World War. Such books are now beginning to enjoy a revival as an end-of-Empire nostalgia becomes fashionable and the new middle class appropriates the history of an old colonial middle class through films like *Out of Africa* and *White Mischief*. The success of Ralph Lauren's fashions and, particularly, his perfume *Safari*, also builds upon this view of a stable colonial Africa.

The 1980s saw a massive increase in the popularity of travel writing, which many people thought had died out with cheaper and easier long-distance travel. Alongside the reissues of the classics of the Victorian and Edwardian eras there was a new wave of travel autobiographies. The latter glorify travel as an individual endeavour in an age of mass tourism and differentiate the traveller, who suffers in order to experience an authentic difference, from the tourist, who just consumes 'home plus'. If the age of scientific description and classification served to underwrite imperialism, the new experiential accounts work to commodify 'abroad' and the foreigners who live there. Serious academics have always sought to distance themselves from both tourists and travellers, but in doing so they are often flattering themselves. Clifford, in his paper 'Travelling cultures' (1992) seeks to reincorporate the sense of travel into ethnography and to reclaim the status of traveller in understanding the contemporary world.

Clearly, however, not all ideas of foreignness are predicated on Orientalism and imperialism. If we think of the phenomenon of international second-home ownership, we can see that the idea of foreignness can be brought to our doorsteps. The very use of the word 'home' introduces a confusion when the house or apartment is situated in a country where one is not domiciled. Quite clearly, foreignness must be seen as a very plastic concept. From the point of view of the owner, the second home grants a vicarious insiderness, often with long-standing relationships with those who live there permanently or with others who visit regularly. On retirement the second home frequently becomes a permanent place of residence, but it is evident that the espousal of 'foreign' lifestyles on a more or less continuous basis is an important part of the attraction. In Britain there is a flourishing genre of writing which deals with the tribulations of those trying innocently, and with poor local cultural skills, to establish their own little niche abroad. (Well-known examples are Bogard's *An Ordinary Man*, 1989, and Mayle's *A Year in Provence*, 1990.) There are obvious parallels with the late imperial mode, with an emphasis on domestic detail and incomprehensibility, the foreign being a mix of authenticity and strangeness, but this time there is an assumption that these qualities are open to appropriation, and are not merely governable. Here the foreign is simply something many people escape into, away from the pressures of the permanent and routinized home.

We can see this growing out of the style of writing which we associate with Bohemian self-discovery through exposure to the foreign, frequently portrayed as a rejection of ascribed lifestyles at home. The ways in which James, Miller, Hemmingway, Fitzgerald or Kerouac write about Paris tell us more about American than French culture, but they also construct some of the most powerful images of Paris that are consumed in the West; images which have been appropriated by subsequent generations of people who seek to cast themselves in the same drama, either as the exuberant plunderer Henry Miller, or as Anais Nin, the resident who found herself simultaneously liberated and exploited in her relationship with him.

We begin to see that the term 'foreign', whether used in relation to places, people, things, ideas, must itself be seen as a part of a process of differentiation, it has no essential meaning and no existence other than in the project of boundary-making and maintaining, it is a term whose function is to exclude, by making strange, from whatever it is that one wishes to reserve as one's own construction of community. If one refuses to see categories and boundaries as anything other than constructions, it becomes possible to see difference as a quality of continuity, a flow rather than a rupture. Breaks in the flow must be seen to be artificial, just like a dyke which separates the land from the sea. Geographers must ask again why boundaries have been put where they have, by whom and for whose benefit. We take this subject up again in Chapter 8.

2

Familiar places . . . home thoughts

There was a song in the musical *Paint your Wagon* which contained the cynical lines:

> Home is made for coming from
> And dreams of going to,
> Which, with any luck,
> Will never come true

It catches the conflicting emotions many of us have about home in a rapidly changing society. We look to home as a place and as a set of ideas which will nurture and reassure us with its eternal verities, then we feel trapped and constrained by it. Many geographers are just the sort of people who are 'born under a wandering star', drawn to the exotic, they are rootless and dissatisfied with the cosiness of the familiar; yet there can be few who do not have dreams of 'home', whether they want them to come true or not.

It is this ambivalence about the concept of home for people in contemporary Western societies that we intend to examine and in this chapter we want to try to tease out the contradictions and the nuances of this value-laden term. We acknowledge at the outset that we might well seem to be two rather strange people to be tackling the idea at all – Kevin with his army background which periodically uprooted his family with little notice, living in Northern Ireland and Germany, going away to boarding school; Pam with her colonial past, coming back 'home' to Britain in her early twenties and finding herself totally lost in an unknown country. But perhaps everyone's angle on 'home' is a little abnormal, perhaps the 'normal' view of home is a construction which is unachievable in reality; perhaps home is always a powerful dream, an archetypal metaphor, mobilized for a whole spectrum of purposes?

Try translating the word 'home' into any of the languages you know. It is likely that you will be stuck for a term which gathers together the same package of ideas as you invest in the word in English. We are searching for something which contains ideas about origins but which can also be used to refer to the place where one rightfully settles – you can leave home to set up a home of your own; it absorbs thoughts about the family (however defined) but also about individual contentment (when we are told to make ourselves feel at home we don't expect that we should immediately switch on a whole spectrum of Oedipal guilt); it can have something to do with houses and land as property or other form of entitlement, but it is possible for one's home to be sold to someone else and it still be home in the sense of the place where one's roots are. The same word can be used to refer to the place where one is currently staying (even on holiday) and for some people 'home is where I hang my hat'.

Estate agents have a tendency to equate the word with 'house', but most buyers do not believe that the 'affordable starter home' is really home until it has been personalized. Homes can be as small as a share of a room, they can be villages, towns and regions or they can be a whole country (and it is certainly not necessary that one has ever been there), but they can also have nothing at all to do with a fixed place – they can be entirely abstract or relational, e.g. one can be at home with a thought, a person, one's own body. It is no wonder that most languages do not have the possibility of a single concept which can contain the same spectrum of ideas about bricks and mortar, kinship, tradition, contentment, regional loyalty, duty, community, nationalism, return, aspiration; why the French *chez soi*, *foyer*, or *toit paternel* come nowhere close to what we are trying to catch. But why do we load so many thoughts and feelings into a single word? Why should we package them up together so that very different notions come to infect one another? What do the various meanings have in common with one another and what elisions are taking place within this polysemy?

Williams (1976) said of 'community' that it was a 'god' word, that is that it is always used positively (even if sometimes the positiveness takes the form of euphemism), and we can see that 'home', which often slides into 'community' has much the same status. One may wring one's hands with despair about the inadequate nature of other people's home backgrounds, but it is seen as a terrible disloyalty or a sign of weakness to do the same about one's own – home should be sacred, and when it is not, then one should remain silent on the subject. Moms and apple pies come in very variable quality, yet the idea should remain untarnished if it is to have any potency. We both have great difficulty with concepts where one is supposed to buy the whole pack simultaneously – they always seem like those auction lots where unsaleable items are slipped in with the attractive ones. And yet most people within mainstream Western culture, particularly those of Anglo-Saxon origin, cannot easily extricate themselves from the power of this idea which rolls up people, place, belonging and stability so very effectively.

Although the idea of home is such a familiar one, there is a remarkable lack of real knowledge about what goes on in other homes, for homes are almost by definition shrouded in secrecy, glorified as a place of retreat from the public realm. (Here we are thinking about home in the sense of a domestic household.) It is this very privacy of the home which allows it to retain its semi-mystical force, for the ideology remains intact if it is not exposed to public scrutiny. Most of us are more familiar with the domestic practices of the characters in soap operas than we are with those of our next-door neighbours, whose privacy we have been trained to respect. If soap operas do not necessarily portray perfect lives, they invariably portray the emotional recreative power of the domestic structure where, after rift, divorced offspring are absorbed back into the parental family, criminal sons are welcomed and rehabilitated, unemployment is understood and built upon creatively, errant teenage daughters and their babies are recognized, joys and sadnesses are shared. Death, sickness and despair are incorporated into life via an unfolding drama with petty denouements. The underlying message is that whatever disruptions and transgressions have occurred, home is the place where, through love, healing can happen and the world be faced again with personal strength. It is very difficult to reject a myth as powerful as this one, especially when it involves warmth, cosiness and familiarity. Houses that look much like the ones 'we' live in (or feel we ought to live in) provide the setting for contemporary morality plays which construct fictitious communities offering spurious 'truths' about 'universal' human relationships. Soap operas employ realist conventions to convey hyperreal experience of realms of human relationship which never existed. We find difficult issues like AIDS, incest and domestic violence incorporated into the soap-opera format, never fundamentally challenging the sentimental power of home – the creed which, if espoused, can reconcile all rupture, not necessarily with a happy ending, but with acceptance and resignation, as the genre weaves a panoply of joys and woes into 'life's rich pageant'. It almost seems churlish to be suspicious of a myth this powerful and seemingly so benign.

So strong is the notion of the oneness of the domestic cycle and the dwelling place that we are capable of involvement in 'home' stories which are way beyond our personal experience of place – American and Australian soap operas are popular throughout the world – and depend more upon an assumption that there are archetypal relations than upon recognizable specifics. They are watched in shanty towns and high-rise blocks and they travel well because of their mythical quality, for their realism rather than for their grounding in reality. However, like all myths, they have an ability to construct and to conserve ways of being. They are morality tales because they form the focus for gossip and evaluation amongst real people in real places, and the lessons about loyalty, duty, continuity, and structure of home are learned as surely as if they were handed down from the pulpit. The soap opera absorbs the world into the construction of home, whereas, in life, home is the place from which one ventures out into the world (Geraghty, 1991). Quite artificially, they restrict the space of social intercourse and impose unrealizable norms of domesticity.

The autobiography is another source of vicarious experience of the privacy of the home. Homes become the ground floor of the edifice which is constructed as a life story. For the story to have meaning, the home life must be presented selectively, seen from the vantage point of the person that the author has become. In his autobiographical play, *A Voyage Round my Father*, Mortimer reveals an eccentric but secure home, dominated by the twin forces of his father's never-mentioned blindness and a cherished garden. The home, place, people and behaviour, is a haven from the outside world, to which he returns in school holidays, on breaks from a cramped marriage and eventually as his own permanent dwelling. The house was not just an inheritance, it became the emblem of a patriarchal continuity which is rare outside agricultural or aristocratic settings, but it is hard to resist the feeling that the author is just a little too pleased with the human being who has been so fortunately constructed. In complete contrast Steedman's *Landscape for a Good Woman* (1986) ought to be required reading for those who extoll the virtues of home. It is a chilling account of a life pared down in loveless resentment and resignation, where saving for a never-purchased house becomes her mother's excuse for a mortgaged childhood of restraint and sense of indebtedness. But Steedman, just as surely as Mortimer, is using her depiction of her home as a device for presenting and validating the person she grew into. She gives only the slightest hints about the place she comes to as a mature woman – that remains her secret space and we do not know whether it is conceived of as a home. This is a trope which is also used by many of the writers in Heron's (1985) collection on girls growing up in Britain in the 1950s – home as a launching pad.

The notion of home is one which has been picked up particularly by that strand of geography which is described as 'humanistic'. An underlying theme in humanistic geography is that of the search for 'authenticity' and a belief (often explicit) that the processes of industrial modernism, with an accompanying growth in commodification, rendered human environments inauthentic as 'man' was progressively separated from 'his' labour, and that the postmodern took in authenticity to even more terrifying proportions. Homes were seen by humanistic geographers, such as Appleton (1975, 1990) as key *places* of experience and identity. They are fond of telling us that the human scale has been forfeited and that 'man' is searching, homeless, for an identity in the contemporary world (Relph, 1976). We do not find it useful to equate alienation with a sense of 'homelessness' or to conflate being happy with one's situation (literal and metaphorical), with knowing one's proper place (literal and metaphorical).

The humanistic trope of being at home in the environment turns upon the Heideggerian notion of *Dasein*, most commonly translated as 'dwelling'; indeed the very word 'environment' should itself be understood as a dwelling, a realm in which one lives. Dwelling and being are very different concepts, the former contains within it the idea of human beings fitting and filling their space physically and metaphysically, it is relational as well as existential, and people, space and material environment are perceived as contributing to one another's identity in a process of mutual 'ownership' (which may or may not have anything to do with property). Closely bound up in this idea of dwelling is that of appropriateness, commonly conceived of as a harmony between the way of living and the land which sustains life. The perfect illustration of this is sought in the idealized lifestyle of the solid peasant, living in a house built from local materials by himself or by his ancestors, wresting a livelihood directly from the soil, making his own furniture and implements and buying only carefully selected essential items from the market. This is a life where the fruits of one's labours return to one's household, where there is a strong attachment to the things and people of the past, where the present does not represent a dislocation. It is also a life in which people know their place, in all senses of the word. Heidegger himself glorified the farmhouse in the Black Forest as the ideal exemplification of his notion of *Dasein* and his followers have offered up their own suggestions or times and places where this harmonious dwelling has been achieved. The trouble is, wherever one is, it always seems to be some time and some place else! We would like to suggest that it is only to be found within the pages of books of fairy stories – where honest woodcutters are rewarded with near-perfect sons as a consequence of wishing. This is not to trivialize this powerful concept, merely to doubt that it has any basis in observable reality, for the most dangerous ideas have their origins in the realm of imagination.

One of the great disappointments of anthropology, for those who do not practice it, has long been that ethnographers have never found the perfect societies where man lives in absolute harmony with his environment; real peasant societies are dispiritingly unharmonious, and it is a gullible observer who assumes that the 'golden age' has only just been lost. Peasants certainly often do have a long family attachment to place, but it is our task to understand why it is that city dwellers have invested this attachment with an overload of sentiment. A recent example of this equation of 'primitives' with a privileged atunement to nature may be found in Pocock (1993).

You may have noticed that in the last few paragraphs we slipped into an uncharacteristic use of the masculine third person – it was not unconscious. We needed to invoke ideas about patriarchal power, whether accepted or rejected. Homes are very differently represented by men and women, wherever the concept exists. If we think about that idealized peasant society for a while, let's say in India, the different status of a man and a woman in relation to home becomes immediately apparent. In an Indian rural environment it is usual for marriages to be arranged by relatives (sometimes via marriage brokers) between young people who come from different neighbourhoods, since clusters of villages form exogamous vicinages. The point is that one should *not* marry someone from 'home', someone too familiar, someone whose family is known too intimately, or familiarity not only breeds contempt, but, more importantly, may result in local political and land conflicts. Women (often painfully young girls, regardless of the legal age of marriage being 18) leave home in the sense of family, friends, village, to join the households of their husband's parents. It is not expected that they will feel at home here, but from now on they will visit their natal home only with the status of visitor and will usually travel there with an escort from their husband's family. In their new environment they will be under the close scrutiny of their mothers-in-law who will instruct them in the ways of the household and guard closely over their modesty and reputation. Sometimes this is done with considerable compassion as the older woman remembers what it felt like to come into the self-same continuing line of men, quite possibly the same house; but quite frequently there is no empathy at all. Almost invariably there is a sense of strangeness on both sides and considerable irritation at exposure to different ways of doing things. Young brides are not just newcomers, they are also outsiders; they have to be incorporated, but the process cannot be achieved overnight. The physical separation from their 'own' families is partially symbolic of the transfer, but it also serves to retain the integrity of both households with their secrets, making it difficult for women to gossip to their mothers about their mothers-in-law. The young couple will be given little time together and it is likely that they will not share a bed but meet clandestinely for intercourse. When the first child is born the girl may well go back to her 'own' home to be attended by her mother, but mother-in-law will rapidly assume a day-to-day supervisory role over the new baby, who is after all entitled to membership of the family by birth, not marriage. If it is a boy he will inherit property and ideally stay on in the home of his ancestors. If it is a girl, the family will have to accumulate the dowry to enable her to marry well and the family to make an honourable alliance. For an Indian peasant man home is a fixity across generations in terms of land, house, patriline, customary behaviour. For a woman it is something she precariously acquires once she is the mother of adult men; there is no place where a long line of female ancestors lovingly tended their land, where domestic lore was handed down uninterruptedly from mother to daughter, or where women's heirlooms accumulated.

The Indian example is similar to that of most peasant societies – Campbell (1964) tells us an even more dramatic story of rupture for Sarakatsani shepherds in Greece, and Foster (1976) shows that in the village he studied in Mexico the loss of a daughter is seen to be so final that the parents claim that the girl must have been raped before permission is grudgingly given for her marriage into another family. Once married, the brides are the most likely candidates for accusations of disloyalty, even witchcraft, when things go wrong. As the collection *Gifts and Poison*, edited by Bailey (1971), amply demonstrates, in one peasant society after another, a good woman is one who is self-effacing and circumspect in her behaviour, gives no cause for gossip, spends as little time as possible in the public realm, but labours diligently within the home, helping accumulate wealth and adding to its reputation. But *whose* home is it? It would be unfair to suggest that women do not come to regard their marital residences as 'home', especially once children are born and when the power of the mother-in-law recedes, but we need to appreciate that their sense of belonging is very differently constituted from that of the men. Women see themselves as *coming to* and *becoming in* the home. In most peasant societies it is men who see these homes in terms of origin, permanence and continuity. Standing (1991) shows how middle-class Indian women in Calcutta are expected to preserve the respectability of their husbands' families by not working outside the home, even though there may be considerable economic pressure upon them to do so and though they may take on poorly paid home-work which can be hidden from the rest of the community. The woman should be installed in her husband's home, with or without her mother-in-law.

Seamon and Mugerauer's collection, *Dwelling, Place and Environment* (1985), is much concerned with the notion of persons at home in the world but is remarkably uncritical about the concept of home, upon which such a heavy load is hung. Several papers in the book make the assumption that authentic dwelling, authentic built structures, authentic social relations, emerge where man and land are attuned to one another and where human practices are not self-consciously devised. Home and traditions elide into one another with considerable sleight of hand and it is implied that people do not feel at home with the new. We want to know why so many (male) geographers want to believe in authenticity that is inherently conservative, for we are firmly of the opinion that people who derive their livelihood from the land and are governed by tradition do not have more authentic lives than those who are workers in factories or hostesses in nightclubs.

Seamon's contribution to the collection is an analysis of the four 'Emigrant' novels of Moberg, which form a fictionalized account of the migration of Swedish peasants to the United States in the nineteenth century. He is concerned with the way in which people from a traditional background experience a problematization of their 'dwelling' in the form of a progressive industrialization of their society; they seek to resolve this by emigration to the New World where they gradually establish a home for themselves. Seamon sees the novels as offering a 'dwelling-journey spiral' where ideas about being in and out of place are resolved.

He is far from gender-blind in his account of the lives of the married couple, Karl Oskar and Kristina, and their adjustment to the colonial task of taming virgin land and establishing a new community. He sees that the active, frontier-pushing endeavour can satisfy the man's desire but only the process of consolidation can offer satisfaction to the woman. Seamon appreciates that the setting up of a new home in a new land offers a man the proof of his virility, his autonomy and his refusal to be mastered by a structure he does not like. How could he not understand this myth? It is one we have all been tutored in by hundreds of 'Westerns', space exploration and adventure movies since early childhood? It is the glorious story of undomesticated man, carving out a 'home', bringing nature under the hand of civilization and all there is for women to do in this story is follow diligently behind their men, providing comfort, making the home homely. Kristina accepts the decision to migrate with considerable misgiving and finds adjustment to the New World painful and slow but, as a community forms up around her and the pioneering recedes into the past, she grows into her new home, incorporating practices and beliefs from Sweden into a new context. When she dies her husband finds that he no longer feels at home in what he colonized and she worked upon, he longs for Sweden and his past and for the first time feels the futility of the endeavour, retiring from working and communal life.

Seamon approvingly quotes the words of Karl Oskar, 'I seek a land where through *my* work I can help myself and *mine*' (Seamon, 1985, p.230, our emphasis), without questioning the patriarchal relationships which underpin such an apparently simple desire. A few pages later, he describes the way in which the man chooses the best land for his farm. 'Feeling comfortably in place for Karl Oskar is largely the completion of the settling process – i.e. satisfying material needs and gaining physical *mastery* over place. Building a house and *moving his family in*, Karl Oskar feels relatively at ease: "Beginning this day, he felt settled in and at home in North America"' (p.234, our emphasis), then goes on to say 'Kristina, in contrast, cannot so easily adjust to the new place'. Seamon interprets Kristina's role in the novels as representing a deeper sense of coming to terms with rupture and newness which she works through for the rest of her family. Kristina is 'homesick' in the new home he has installed her in and she seeks comfort in the familiar things she has brought from Sweden. Although Seamon's analysis is sympathetic in *human* terms, he misses the simple interpretation that a woman would put upon the story; Karl Oskar feels *at home* in that Heideggerian sense of dwelling because of his mastery. 'Moving his family in' – how could Seamon write such words if they were politically neutral? How could he not feel for Kristina, *moved in*, not actively *moving*?

Seamon chose a particularly powerful version of the 'dwelling' myth in that the quartet draws upon ideas about traditional peasantry and the frontier, tradition and noble innovation. The construction of the male hero would have been different if the family had come from an urban background or moved from peasantry to industry but it would still have revolved around mastery, achieved or frustrated. The woman's role in this drama of home construction is one of struggle for the good of the rest of the family, of acceptance of the endeavour and smoothing the way until she finally looks back and sees a new woman, different from the one who left home. In Chapter 10 we look at different ways in which the changes over a life course can be conceptualized, following the idea propounded by Deleuze and Guattari (1987) that there are three major ways of thinking about linearity: (1) the line of rigid segmentation, where episodes can be marked by clear breaks; (2) the line of supple segmentation, where tiny cracks accumulate imperceptibly but then threaten a fissure; and (3) the line of flight – a rupture which allows no turning back and allows little to be carried forward. We can map the different responses of Karl Oskar and Kristina onto this surface: Karl Oskar following the line of rigid segmentation and then the line of flight; Kristina the line of supple segmentation.

But Seamon interprets the story according to a universal theme based on the 'rest–movement relationship' which he sees as the '. . . associated polarities of home and reach, centre and horizon, dwelling and journey'.

> Rest, the opposite of movement and journey, relates to a basic human need for spatial and environmental order and familiarity. Rest anchors the present and the future in the past and maintains an experiential and historical continuity. From the vantage point of human experience, the deepest manifestation of rest is *dwelling*, which involves a lifestyle of regularity, repetition and cyclicity all grounded in an atmosphere of care and concern for places, things and people. (Seamon, 1985, p.227)

It all sounds so *lovely*, until you question whether the past really was so wonderful, whether continuation of the structures you live in is to be preferred to a liberating change, whether 'care and concern' are based on your, or someone else's, notion of 'good'. Dwelling is inherently *conservative* and there is nothing wrong with that if you are sure that you do want to conserve – we are not, and we feel as oppressed as Nora in her 'Dolls House' when we read Seamon's view that 'one's personal and familial situation must be stabilised and ordered before a sense of extended community can be re-established' (p.240), given that neither of us resides in the position of the patriarch.

We are depressed that the notion of home seems invariably to be ranged in battle against change or even motion. Why is home so frequently depicted as *defensive* and inward-looking? Why has Appleton's (1975) view of *refuge* gained such currency? Why do people use the language of warfare at all when contemplating the domestic? In Chapter 7 we explore some of the ideas about the public/domestic opposition and the way in which it takes on different meanings for married women and men. In Pam's family is a plate, given to her parents as a wedding present during the war, a cherished object, evoking childhood crumpets in the winter, eaten, not in the country cottage depicted on it, but in an Essex council house. The picture on the plate is of a woman sitting by a fire, darning a sock. The kettle is ready for tea and a larger more comfortable chair than hers stands empty, facing her. Beneath the picture is the legend 'An Englishman's fireside'. It is only relatively recently that it dawned that there was any room for cynicism about this particular image of warmth and nurture, home and family. Light's *Forever England* (1991), a study of conservative women's literature between the wars, tackles this question of respectable resignation to domesticity as a suppression of desire. She quotes Delafield's *The Way Things Are*, a story of a woman's struggle between her horror of her circumscribed domestic role and her fear of rejecting it: 'It dawned upon her dimly that only by envisaging and accepting her own limitations, could she endure the limitations of her surroundings' (Light, 1991, p.140) – the editing of a woman to fit the boundaries of the segments she inhabits.

It isn't firesides, love and security we are opposing, it is 'dwelling', with all its acceptance and resignation, fixity and rigidity, its boundaries and exclusions, its retrospectiveness and its glorification of folksiness, order and conformity, and the implicit surveillance, patrolling, gossip and punishment for transgression. 'Dwelling' conjures up for us an environment of carefully kept accounts of rights and obligations that spells a lack of generosity and *jouissance*. Home only has any meaning in opposition to 'not home', the place where the outsiders reside and from which one retreats; the very concept of home assumes that there are enemies beyond who may not be admitted into the private realm. Generous people keep 'open house', no one keeps an 'open home' – the very term is a contradiction.

Süskind's *The Pigeon* (1988) catches this sense of selfish, closed in possession, in a novel of creepy obsessiveness. Noel, the protagonist, has lived for years in the same small room. It is modest but immaculately clean and everything he requires for his pared-down life is contained within it, neatly put in its own proper place. One morning he opens his door and finds a pigeon trapped on the landing. A feather and a splodge of bird-droppings are on the floor. He is sickened and feels defiled and is thrown into a panic which infects his whole day, jeopardizing his job, his self-esteem and his ability to cope. One tiny thing has made his home alien and he cannot bring himself to return to it until he has been through a nightmare of self-debasement, which takes him to an enlightenment that it is not the dirt of the pigeon, but the enveloping safeness of the home which has caused his extreme reaction to the invasion of disorder. His alienation from his home causes him to hate all his fellow humans, seeing them defiling the city: 'Filthy pigs! Hooligans! They ought to wipe you out. Yes they should! Flog you to death and get rid of you. Shoot you down. Every single one of you all at one time' (Süskind, 1988, p.61). This reaction to the exhaust fumes of cars in the language of the 'final solution' encapsulates Süskind's depiction of the slide from meek orderliness to fascist obsession with control and it is not far fetched to read the novel as an allegory for European politics.

We have little time for those who seek to discredit philosophical positions by referring snidely to the political allegiances of their main proponents or, even worse, the political activities of people who appropriated their ideas, but we believe that the concept of *Dasein* itself is *inherently* fascist, whether one is looking at the great fascisms or the little personal ones Deleuze and Guattari alert us to. It ties together so many structures – spatial, political, gender, kinship, economic, emotional – that its deconstruction must be a matter of a severence. It is no wonder that there is a literary genre which charts the escape from the ties of home to balance that which glorifies it.

The literary genre of *Bildungsroman* charts the journey away from home to seek one's true identity in experience. It is the story of the journey from youth to maturity depicted as a quest; the underlying assumption in such fictions is that a boy can only grow into a man by detaching himself from the site of his childhood, particularly from his mother. Freed from routine and the familiar, a romantic heroism is constructed in the context of a willed self-discovery. The home is a point of no return and the wide world lies outside, harbouring one's destiny. This is a peculiarly masculine genre, often involving seafaring to exotic locations as in the works of Conrad (1898, 1975) or Lowry (1933) where home is depicted as soft, feminine and lacking in adventure, insufficiently testing.

Conrad's *Youth* tells the tale of the making of a young man on a voyage from Liverpool to Bangkok; the ship is unsafe to start with, the cargo catches fire and explodes, but still the journey continues in its glorification of untrammelled heroism. This male identity is constructed not just in opposition to the feminine (in the form of the fussing and overprotective captain's wife, worrying lest her husband forget to put on his muffler as she waves him goodbye on a voyage which brings him close to death), but also with the sea in contrast to land, the East in contrast to the West. In all of these oppositions the latter part is associated with the constraints of home, the place of compromise where a man's true spirit cannot be realized. At the end of the story, told by a successful middle-aged man, long in a professional shore-bound occupation, there is regret for the passing of youth and the compromises made by the solid householder:

> By all that's wonderful it is the sea, I believe, the sea itself – or is it youth alone? Who can tell? But you here – you all had something out of life: money, love – whatever one gets on shore – and tell me, wasn't that the best time, that time when we were young at sea; young and had nothing, on the sea that

> gives you nothing, except hard knocks – and sometimes a chance to feel your strength. (Conrad, 1898, 1975, p.39)

This is a far cry from the homely securities of *Dasein*, where the status of the middle-aged man, settled and surrounded by his family, is glorified as the proper aspiration of masculine power.

There are also the many novels which we can see as representing escape attempts, Maugham's *The Moon and Sixpence* (1919), for example, has a man throwing up a secure job, wife, family and respectability to seek the lifestyle of an artist far away from home. Trollope's *The Rector's Wife* (1991) similarly depicts home as restraining in its demands for structured living and has the protagonist leaving home, first just on a daily basis to do paid menial work which is not regarded as sufficiently respectable for a clergyman's wife, then making a decision to walk out of her marriage and the church. Here home is not the secure starting point for an adventure but the negation of real life; it is presented as a site of meaningless restrictions or an emotional void from which one must run, not in search of adventure but merely in order to live fully. The role of the paterfamilias is worked out as a hollow sham, whether for its incumbent or for others.

Yet another anti-home device in literature and film is that of the uncanny (in German *unheimlich*, literally meaning 'unhomely'). Here the home, as a building, takes on a malevolent character and seems to be destructive of its residents. A notable example of this is Du Maurier's *Rebecca* (1981), where the house, Mandalay, and its housekeeper reject the new wife of the householder, who cannot feel at home with the place or the memory of its former mistress. The house has to burn down before there is a chance that she will feel at one with her husband. Or we can think of the film, *The Haunting*, so ably deconstructed by White (1992), where the house has the power to destroy its residents. Even Brontë's *Jane Eyre* (1964) requires the burning of the house, the death of its evil spirit in the form of the alien and mad wife, the blinding and enfeeblement of its patriarch, before Jane, homeless all her life, can build a marriage.

All these examples have, as a central character, an individual, repressed or threatened by a man's home, struggling to be free of it in order to be at peace. It is possible to interpret these as a rejection of static residence and ascribed relationships as repressive ideals; as such, they are a glorification of boundlessness, mobility and insecurity. Much of what is being rejected is a politics of dwelling.

The political aspect of the concept of dwelling is most apparent at the level of ethnicity and its intersection with nationalism. The idea that 'we' groups should have their own home is appealing until one considers what happens to the 'they' groups that have to be excluded. The notorious programmes of 'ethnic cleansing' in Bosnia, the forcible construction of 'Bantu homelands' in South Africa under apartheid, the genocide of

Amazonian peoples in the cause of the Brazilian frontier, Sikh terrorism in the Punjab, are all recent examples of the power of the concept of *Dasein*, the (masculine) desire to be oneself in one's own place. We will look at this in more detail in Chapter 8.

The benign word 'home' can be used euphemistically for something which has no sense of dwelling, but rather an incarceration, set up, like the Bantu homelands, for the purpose of excluding 'marginals' from the rest of society. Old people's homes, mother-and-baby homes, nursing homes, children's homes, homes for the mentally handicapped, homes of the blind, etc., where 'residents' are rigidly distinguished from staff, and there is no suggestion that the residents are dwelling at home and the staff are their servants. We find that in such places brochures may well say such things as 'residents are permitted (or encouraged) to ...' bring items of their own furniture, receive guests at approved times, have their own telephone lines, go out on visits – precisely the sort of things which, so long as you could afford them, you'd need no permission or encouragement to do, normal things which in their control can be the structuring of abnormality in those marginalized places where surveillance is the rule. Of course prisons, schools and hospitals are never called homes – in these inmates, students and patients submit to a regime, they do not connive at it. In institutions called homes there is at least a pretence of voluntarism, that people are living there 'for their own good', abdicating major decision-making 'in their own interest'. No one ever talks about 'living in' a prison, school or hospital, just being in it.

Kidder's 'The last place on earth' (1993) is a moving account of the micro-geography of extreme old age in a 'home'. She fictionalizes her ethnographic experiences to tell the story of a death and two 'survivors' in a setting where men attempt to control small tracts of personal space, or manipulate corridors in order to prove that they are capable of escaping into semi-public space. Mobility looms large in the three men's sense of personal dignity. The outside world takes on an unreal quality, news and visitors come in, and telephones appear like emotional lifelines, but when the residents do go out they find it more exhausting and bewildering than stimulating. They have shifted their horizons to their living space. We, and the residents, know that, though they can never go home, they must make themselves at home if they are to survive.

Would it be fair to say that there is no concept of home without some sense of loss, whether experienced or impending? As children we learned the maudlin Cockney drinking song:

> Show me the way to go home
> I'm tired and I want to go to bed.
> I had a little drink about an hour ago
> And its gone right to my head.
> No matter where I roam,
> By land or sea or foam,

You will always hear me singing this song
Show me the way to go home.

A song of terrifying pathos which switches from the locally to the globally to the existentially lost in so few lines, it was intended to warn against the ruin associated with drunkenness. Yet, for many people there can be no home. We deal with the subject of homelessness in Chapter 9, but for the moment we should also recognize that:

> the security of the boundries of the place one called home must have disappeared long ago, and the coherence of one's local culture must long ago have been under threat, in those parts of the world where the majority of its population lives. In those parts of the world, it is centuries now since time and distance provided much protective insulation from outside. (Massey, 1992, p.10)

If one's home, conceived of as a refuge, is invaded or appropriated it ceases to have the same value; it becomes a site of nostalgia, which properly means 'homesickness', not just a longing for the past. Doreen Massey quotes bell hooks on estrangement:

> home is no longer just one place. It is locations. Home is that place which enables and promotes varied and every changing perspectives, a place where one discovers new ways of seeing reality, frontiers of difference. One confronts and accepts dispersal and fragmentation as part of the constructions of a new world order that reveals more fully where we are, who we can become. (hooks, 1991, p.149)

Home is more importantly a point of departure than a place to return to (see Box I).

BOX I From *Never Go Back*

Outside, the streets tear litter in their thin hands,
a tired wind whistles through the blackened stumps of houses
at a limping dog. *God this is an awful place*
says the friend, the alcoholic, whose head is a negative
of itself. You listen and nod, bereaved. Baby,
what you owe this place is unpayable
in the only currency you have. So drink up. Shut up,
then get them in again. Again. And never go back.

(Duffy, 1993, p.30)

This sentiment is captured vividly by Wolfe in his novel *You Can't Go*

Home Again (1940, 1962), where the protagonist finds that, having made his reputation as a writer with a 'home-town' novel, not only is he unacceptable in the town he describes, he also feels a disgust at the way it is writing its own future in speculative land deals. The book then proceeds through a series of living places, all relating to facets of his known Self, none of which he can feel at home with, until on a lecture tour he meets his ideal woman whom he is obliged to leave in Germany when he flees back to America as war is about to be declared. He is once again homeless in his home country.

In similar vein, we find a genre of writing about the British colonial experience which focuses not on the masculine appropriation of Empire but on women's seemingly disloyal inability to easily come to terms with the place they are expected to regard as home. Madness or moral frailty are often used as the literary device for representing this homelessness/hopelessness. Lessing's (1953, 1964) fictionalized accounts of her childhood in Rhodesia show her mother, living on a desperately failing farm, resentful, closed in upon herself, snobbishly inventing a refined 'past' at home in England, unaccepting of the world of Africans and Africaaners around her, buttoned up in her 'Englishness', which teeters on the edge of madness. This type is also familiar in the writing of Somerset Maugham, Graham Greene and many others who have examined the colonial experience. What is fascinating is the way in which unmarried women readers often seem to identify with the men in these stories, the men who dream of mastery seem infinitely desirable, as does the promise of a home constructed especially for oneself out of alien territory. The women playing out the colonial role are so often portrayed as petulant, ungrateful and uncooperative in their discontent, we might ask why.

Yet even this can be constructed as one's familiar base. Lessing's *Going Home* (1957) is an account of a visit back to Rhodesia, where she grew up. On her return on holiday from London to the place where the wide skies and the empty High Veld seemed part of her very being, she was hounded for her political sympathies and declared a prohibited immigrant. The book is now a historical curiosity (Rhodesia has become Zimbabwe, its white supremacy is long since defeated) and yet it still catches the yearning of the writer for the place she loves. It is a place she can only exile herself from, loathing the racist politics and its conservatism verging on fascism. But it is not only the physical environment for which she has nostalgia, the experience of the whole package of rejected white racism, which has constructed her starting point, is also a homecoming, but one which is another point of departure; departure into knowing that she can never go home again.

Here we are looking at the condition of the individual who knows that her own past and home is a stigmatized one and yet she cannot reject it because in doing so she would reject her own childhood and youth, that is, herself. Poewe (1988) writes of her wartime childhood in Germany and the way in which, moving to apparent security in America with her family, she can expect no understanding for her story or her self, constructed out of fear during air raids, the food shortages or her mother's prostitution to British soldiers. Of women like her mother she says: 'They live on the periphery or in the limbo of this world. Uncomprehending. Bitter. Defensive. Dumb. They are not intellectuals and cannot weave themselves pretty justifications. No one hears their explanations ... How can you speak of your suffering when your kind caused others to suffer?' (p.119). And of herself she writes: Why guilt? Because, of course, I rejected my birth. I rejected it because it was too heavy to bear. So much did I feel its weight that I came to see my ethnic origins as the major hindrance to my sole ambition, not to have had ethnic origins (p.210). Her book is a reclamation of her lost home, but one which involves no sense of return. Poewe became a social anthropologist, looking for her sense of belonging on a global scale, particularly in central and southern Africa.

Duras expresses the same irretrievable loss in *The Lover* (1985), having exiled herself from her colonial status in Indo-China (Vietnam) and her unacceptable relationship with a Chinese lover, she feels oppressed by the bourgeois values of France. Similarly, Cixous (1979, 1989) in a piece of poetic wordplay, muses upon her space/time/political separation from (French colonial) Algeria. She articulates her vicarious belonging in 'strange' places and her sense of home which, like that of many people whose formative years were spent in an alien place, can only be snatched at. News from Iran starts her meditating on an orange, a symbol of her sense of herself, familiar but exotic, but also a symbol of the world:

> And so, what have I in common with Iran? Nothing, other than a syllable which I can't ignore because it has the power to pull me by the ears back to Oran, the town where I was born. An error, a coincidence but it makes me fall back in line. ... As soon as I leave the old lands an Iran telephones me. When I acquit myself, the debt tracks me down. (p.33, Pam's translation)

Loss, longing, disorientation, sometimes focusing upon the past, sometimes upon the future. Like Cixous' orange, home is good to meditate upon, but we would suggest that it is a slippery subject to build a major strand of human geography around.

Because we have families we love and places we are deeply attached to, we feel that it is important to end this chapter spelling our personal intentions out. What we have written is an acknowledgement that overwhelming emotions can be harnessed for purposes of mobilization and control. When too many sentiments are rolled up into one catch-all term they need teasing apart rather than swallowing whole, lest something very nasty indeed is ingested.

3

Past places

Places in the Western world are socially constructed with a considerable intensity of nostalgia as consciously and unconsciously we create and recreate them with a sense of history. History is not just the traces of the past, but is the outcome of a dialogue between the present and the past; the present itself being many-voiced. As we do this there is a tendency to narrate one history as a prior, dominant history around which supposedly lesser histories compete, and it is this dominant version of the past that will be preserved and represented in physical form, in education and text. What survives intact, is refurbished, or recreated in replica, emerges, in particular locations, from the will of human beings of different powers. Different social groups strive for their interpretation of the past to be re-territorialized in texts of all kinds, including landscapes, 'not for disinterested reasons, but to help them get what they want or keep what they have got', namely access to economic and cultural capital (Thrift, 1989b, p.15). But, for those on the margins, this may be articulated only through a desire to make sense of the world in such a way that they cease to be peripheralized into a supporting role, and appear as legitimate actors in their own right.

Heidegger (1962), building upon Nietzsche's *The Use and Abuse of History* (1980), placed considerable importance upon the role of historiography (the way history is written) in constructing present existence of both individuals and peoples. Nietzsche saw three types of historiography: the monumental (offering great inspirational examples from the past); the antiquarian (involving a reverence for the past and a desire to conserve it); and the critical (which sought to judge and obliterate the past). He saw all these as future-directed, the present being discounted as nothing other than banality. Heidegger took on much of this but saw the three types of historiography as aspects of the same thing assuming, however, that the state of current being would make one more meaningful than the others in terms of directedness, mood and feeling. So, for him, monumentalism would predominate where there was a conscious striving towards the future – here heroic figures and great events would be invoked from the past and the intention would be Utopian. Antiquarianism would predominate where a heritage was revered and conservative behaviours were emphasized, but it would be impossible to take a monumentalist stance without simultaneously being antiquarian. For Heidegger critical historiography was directed not at the destruction of a despised past but at the destructuring of the way that the present looks at its past. Later, Heidegger (1980, p.155) stated unequivocally that 'history ... if it is anything at all, is mythology'. We take these views as axiomatic as we look at history as it is configured in place, but we also want to problematize the monumental and the antiquarian by thinking critically – which heroes, which great events inspire and depress which people? Which pasts do we revere and which render us abject? Heidegger saw that questions of heritage and destiny are inextricably linked and that this is why heritage is such an important site of contestation. His view that the project of being is never accomplished until death, and that being and doing should be seen together, implies that the way we construct our past is an inherently political matter. In Heidegger's case, that politics assumed charismatic 'god' figures capable of appropriating the common past and seizing the future; he despised the ordinary people, but revered the people conceived of as 'the Volk', the nation, with a common heritage which would choose its hero – and, because such views are thinkable, it is very important that we do not regard the construction of history as a trivial matter.

Local configurations of power cause a certain sense of time to be written onto places, always to some extent destroying, blurring, retrieving, incorporating and reworking many traces of other times. Places may become associated with certain eras or particular events, to the extent that, in part, the present identity is forged in conscious interaction with the past, in the same way as members of lineages draw much of their own identity from their ancestors. So, for example, Florence is inscribed as a Renaissance place, though it (including its past in the present) clearly exists in dialogue with other contemporary places. The past is not a fixed thing, not just for the obvious reason that the present is constantly slipping into it, but also because the present, itself in movement, keeps on constructing the preferred past. When we say that Florence is identified with the Renaissance, we mean that it is identified with today's appropriation of the Renaissance because it appears in a great many places. However, this appropriation can be interpreted very differently by various residents of the city and by those who incorporate it into an aesthetic or travel experience.

Augé (1993) has drawn attention to the way in which history is accelerating in a multiplication of known world events. By this he means that, given the increasing speed of both telecommunications and physical communications, what is deemed to be a historical, as opposed to a contemporary, explanation of the present draws upon ever more time-close, though perhaps space-distant happenings. (He is implicitly building upon Paul Virilio's, 1984 ideas about speed as the constructing and limiting principle of the late twentieth century, which we explore in Chapter 13.) One of the consequences of this for places is that the inscription of recent occurrences becomes historicized, rather than a part of the present – nostalgic value is accorded rapidly to sites of political spectacle. Spectacles are themselves fleeting and mediatized (for example, the televised drama of the demolition of the Berlin Wall as a ritual of German reunification), but the site becomes a place, capturing a moment in flight. The place is instantaneously communicated around the globe, causing repercussions, which are themselves subject to the same processes.

Enduring place myths are, then, susceptible to revision. Place meanings are always being radically and subtly reterritorialized through a reworking of their past. The example of the Berlin Wall can be seen as an attempt to rewrite a postwar past in order to reunite its two separate histories, but the city had already sanitized its image as the capital of Nazi Germany by looking deeper into its past whilst simultaneously stressing its modernity. But, if history can be used in reconciliation, it can also be employed in conflict as real or imaginary heritages are claimed and refashioned for the present.

Clearly the past interacts with constructions of the present in physical places, but we can also see that ideas and sentiments can be displaced, for purposes of contemplation, into other times and other places. In Chapter 4 we focus upon this fiction in imaginary futures and parallel worlds, but we can see that imaginary worlds are often set in the past too. For example, for many people eighteenth-century Scotland has become a site for romantic patriotism; the American Frontier of the nineteenth century an embodiment of manly independence; Berlin between the wars representative of decadence; East Africa in the 1950s a scene of spartan elegance – one has only to look at product advertising, film location and tourist promotion to find these and many more time place clichés (see Box J).

BOX J *Hunting Mr Heartbreak*

Raban (1990, pp.97–8) beautifully catches the way in which places are often imagined through a pastiche of known and recreated events, ideas and commodities in his description of Ralph Lauren merchandising in Macy's, New York

> There was no agriculture in Ralph Lauren country; no tenant farmers, no peasantry. Industry was restricted to game-

Cont.

Box J cont.

conservation, domestic service and taxidermy. Some ancient system of enclosures had turned the country into a gigantic stretch of parkland. Like Yellowstone, it was conceived on a gigantic scale, but its vegetation and wildlife were, broadly Kentish, with the odd Yorkshire grouse moor and Scottish deer forest thrown in for good measure, along with the polo pitches of Jaipur. . . .

The essential symbolism of this pastoral vision was not drawn directly from England but from the P.B.S. Television – from 'Masterpiece Theatre', and its repertory of soap operas with instructions by Alistair Cooke. Lauren had built his world out of the set designs for Upstairs Downstairs, Brideshead Revisited and The Jewel in the Crown ... Lauren had succeeded in eradicating the last trace of historical reality from his version. He had turned it into a distinctively American fiction as true as a tale of Cockaigne.

Jameson (1985) drew our attention to the postmodern film device of employing particular settings to evoke the attitudes and feelings associated with the historical period that is attached to the place in the popular imagination. He was referring to the use of Florida in the film *Body Heat* as a device for evoking the 1930s, though the film is set in the present:

> there is a faintly archaic feel to all this. The spectator begins to wonder why this story, which could have been set anywhere, is set in a small Florida town. ... One begins to realise after a while that the small town setting allows the film to do without most of the signals and references to the contemporary world, with consumer society – the appliances and artifacts, the high rises, the object world of late capitalism. Technically, then, its object (its cars for instance) are 1980s products, but everything in the film conspires to blur that immediate contemporary reference and make it possible to receive this as nostalgia work – as a narrative set in some nostalgic past, an eternal '30s, say, beyond history. It seems to me exceedingly symptomatic to find the very style of nostalgia films invading and colonizing even those movies today which have contemporary settings: as though for some reason we were unable to focus on our own present. (p.117)

Many film directors use social spatialization and place myths associated with the past to convey traditional or pseudo-traditional values. In both Britain and America, the small-town setting of the 1950s or 60s is often used to signify values of community and kinship, supposedly lost in the consumer society of late capitalism. However, in postmodern cinema, past and present objects, values and times become interwoven. In Steven Speilberg's *Back to the Future* film series, for instance, past events quickly become mixed up with objects from the present, the Delorean car, used as a time machine, being the most tangible example. Despite the car, all of the action of the film takes place in one small local community which stands for the threatened middle-class suburban values of many Americans. Films like this nurse nostalgic desires for past commodity relations of symbolic exchange. *Back to the Future* humorously parodies the full power of Oedipal relations, as the hero goes back in time to find a young version of his mother falling in love with him as a generation is transgressed, though age is not. As in all time travel, the past, present and future are conflated and the result is to construct an overarching experiential sameness in one place. In order to have an existence he has to switch his mother's affections from himself to his real father-to-be, accepted as simultaneously senior, subordinate and same.

We can see striking similarities between the film devices we have just referred to and the themed museum where one is invited to 'step back into the past'. In most affluent countries today, there is a proliferating heritage industry which functions to commoditize the past within the spheres of both tourism and education and acts, commercially, under public subsidy or charitable support, to create place identities. Despite remaining trapped in their own idealism and romanticism in their outright rejection of much of the heritage industry, Wright (1985) and Hewison (1987) have alerted us to the stultifying effects of this apparently benign interest in and presentation of the past in the popular new generation of commercial museums in Britain. These museums often focus upon the way people lived and worked in the past, they are highly localized in their content and accessible in their presentation, using mediatized displays, actors and 'hands-on' experiences. The problems emerge from the association of present places solely with their historic roles, ignoring their present economic decline; a 'prettified' or dramatized view of relations of production in the past; a view of the past as static, rather than processual – an inheritance rather than a continuity. Many museums can also be criticized for their emphasis upon the presentation of the past as purchasable, both as a leisure experience and in reproduction. This latter involves the production of kitsch, which is not always locally or time-specific, for sale in museum shops. These historicized commodities can also become associated with particular gentrified and service-class lifestyles which encourage a range of nostalgic geographies promoting further rounds of heritage tourism. As Urry (1990, p.106) argues, 'the proportion of the service class visiting museum and heritage centres in any year is about three times that of manual workers'.

Museums, are clearly far from neutral, they display certain forms of knowledge as cultural capital and this is often used to reinforce distinctions between different classes, genders, sexualities and ethnic groups. As Porter (1987) shows, women are rarely shown in anything more than a subservient position, either in the domestic space of the kitchen, or doing suitably 'ladylike' tasks in the drawing room. Departures from the male norm are not regularly placed on the literary trail either, with notable exceptions such as the Brontë museum at Haworth and the designation of an area of Wearside as 'Catherine Cookson Country'. (A decision not to save Virginia Woolf's house from being demolished to make way for a planned rubbish dump suggests an insult in its reluctance to allow a major woman writer to be inscribed upon the landscape.) Women, children and ethnic minority groups frequently appear in museums only in supporting roles to the heroes at the centre (those who 'made history'), though there are separate museums of childhood and very rare museums of women's history. Those attempts to depict subaltern histories all too often flounder on the 'speaking for' that we examine in Chapter 8; and at worst they alienate, trivialize and render curious. Without sensitivity they can homogenize where there is variation, they reinforce a sense of the Other, often by emphasizing things which, torn out of context, represent the West's notion of primitiveness and they appropriate in the sense that they translate, rather than reveal.

But museums, like all other texts, have the capability of being read differently from the intentions of their curators or owners. Other histories can be recovered in both the displays and in what is missing. In her analysis of the Canadian Museum of Civilization, Delaney (1991) has shown how the performances of the museum and the individual interact in much more subtle ways than a lot of critics allow for. Museums, it must be remembered, have now begun to focus on the use of exhibits, and visitors have been able to reorganize, and sometimes misinterpret, the dominant narrative put forward by the museum.

There has been a movement to focus upon and question the boundaries between Selves and Others in displays which emphasize hybridity, breaking down the simple view that societies, as bounded groups, 'owned' cultures, as ethnic identifiers. The attempt is one which has the effect of making visitors think transgressively about systems of classification, but rarely can a display do more than snatch at superficials of relatedness and conflict. Hybridity should demonstrate that two sides of a dyad define each other through their contact, but this is a lot to ask from the placement of material artefacts and all too often exhibitions seem to take a dialectical approach, giving the impression that there are two coherent wholes with an interface of syncretism and borrowing. Coombes (1992, p.42), a museum curator, addresses the problem of adequately representing meaning, rather than just form when she writes:

> One of the problems with any exhibition which foregrounds hybridity is that while it may recognise and celebrate the polysemic nature of the objects on display, it often disavows the complexity of the ways in which this is articulated across a series of relations at the level of the social, not only in the culture of origin but also in the dominant culture of the host institution.

Contested histories cannot easily be depicted, particular viewpoints emerge and others lie hidden and curators have to be prepared for the criticism which will almost invariably be levelled at them.

Sometimes this criticism will be passionate. In 1985, for example, the organization Survival International mounted a campaign against an exhibition at the British Museum in London called *Hidden People of the Amazon*, on the grounds that it trivialized the plight of Indian peoples by depicting a smooth transition from indigenous culture to Western culture, ending with a huge photograph of a Panare Indian astride a motorbike, giving the impression of the benefits of absorption into the market economy. Survival International brought Indians to London and orchestrated considerable media coverage, its director being quoted as saying 'it is rather like mounting an exhibition on the Jewish people in 1945 and making no reference to Auschwitz'. The criticism was not so much about the artefacts displayed, as about the story that was not being told. The organization maintained that the Amazon as a place was being falsified in its London representation. A review by Harris and Gow (1985) drew attention to the way in which the exhibition, which they saw as having merits in terms of its depiction of long-house life, fell down in its treatment of the process of change, treating it under the themes 'Discovery, Exploration, Anthropologists and The Clearing of the Forest' – firmly placing Amazonian history in the context of outsider enterprise. The journal *Anthropology Today* carried a surprising article by its assistant editor Houtman (1985) claiming that Survival International was more interested in using the exhibition as a device for getting public support of its cause in aid of threatened peoples than it was about the display itself. The argument revolved particularly around the superior knowledge of the anthropological profession over the charity on matters of social structure and change. The ensuing controversy ran for several issues of the journal, raising questions about the right of the museum to own the objects displayed in the first place, the different responses of differently motivated Indians and anthropologists, and the very matter of whether anthropology as a discipline should have any links with museum history (Leach, 1985; Moser, 1985). The matter of seeing oneself as an Other is not an insignificant one, but here we find that the main controversy revolved around the right of different Western experts to do the representing of the place, the people and its history. (Interestingly, nobody commented upon the male heroic stance of the adventurers involved in the collecting and depicting, or on the fact that all of the Indians consulted for their views seemed to be men.)

A few years later there was an apparently similar case, this time focusing, not on the message constructed through the display, but on the funding and timing of a special exhibition in Canada. As part of an arts festival to accompany the 1988 Winter Olympics at Calgary, Shell Oil Canada Limited undertook the sponsorship of an exhibition entitled *The Spirit Sings: Artistic Traditions of Canada's First People.* Antique art works were assembled from foreign museums as well as those in Canada in a glorification of Indian history. People of the Lubicon Lake Cree were at the time pursuing a land claim against Shell Oil, which was drilling in their traditional lands, and mounted a campaign to stop the exhibition which they saw as an attempt on the part of the company to whitewash its relations with Indian peoples. Indian visitors were divided, some drawing inspiration from the celebration of their past in beautiful artefacts, others resenting the appropriation, exoticism and the sponsorship. The museum and anthropology worlds were divided too; a minority of museums responding to the Lubicon request that they should not lend artefacts. The Glenbow Museum in Calgary defended its staging the exhibition on cultural and expedient grounds:

> Lubicon supporters claimed that Glenbow had accepted money from an oil company which had drilled on land claimed as part of the traditional land of the band. In this era of declining government support, cultural institutions (including universities) have no option but to seek outside support for projects they undertake. This does not mean that corporate sponsors play editorial roles in the theme of the projects that they fund. Nor is there any evidence that the public confuses corporate support for a museum as a museum's support for corporate policy. (Harrison, 1988, p.8)

This view was refuted by a curator at McCord, one of the boycotting museums:

> What was at stake was whether McCord aligned itself on this issue with Native People or with governments, corporate wealth and the glamour of the Winter Olympics. Nor do I believe that museums can accept money from corporate sponsors and pretend to maintain their academic freedom. How many exhibitions do we see that portray corporate sponsors in a critical light? The sufferings that have been inflicted on the Lubicon people by the Alberta and federal Canadian governments and by the oil companies that exploit their land have brought international shame upon Canada and are a warning to all of us about how ruthless governments and big business can become when they are not held to account by a vigilant public. (Trigger, 1988, p.9)

These examples amply demonstrate the area of contestation not only of the representation of pasts, but also the impact that these have upon present uses and appropriation of territory by different interest groups.

However, there are also people's museums where local people have donated their family keepsakes and treasures in order to pool their own history for the benefit of future generations – these are often moving in their intimacy and their personalization of major events. For example, a museum of mining and pit villages in the Avon Valley in South Wales was the work of local people who, with development grants, employed curators to help them display their own version of the past for themselves. The poverty and powerlessness comes through but also the tremendous spirit of resistance, underwritten by the fact that the museum came into existence at exactly the same time as the last pit in the valley closed and people were fearful that their communal identity would be lost. But clearly it was also a part of the wave of new industrial-heritage museums, designed in part to bring employment to a declining area through tourist provision. The museum is as biased as any other, the mine-owners are not glamorized in their lifestyle nor praised for their philanthropy; technological change is not depicted as achievement. It is a partial view of the past, but women and children are there as surely as the mining men and there is a small display of news clippings about a party of Africans, sponsored by the National Union of Mineworkers, visiting the valley in the 1950s, which dwells on the class solidarities and mutual curiosity in the encounter. In the increasing number of such museums, it is cynical to focus one's attention on the past as a commodity as one overhears elderly people using the exhibits to spark off reminiscences to their grandchildren about their childhoods and the stories they were told by their own grandparents. History is brought through into the present and can have an important part to play in establishing a commonality of spirit between local people.

Many museums today are laid out in such a way that a set path must be taken through them; the activity of viewing may be heightened by a surrender of the observer's powers of locomotion to a ride past displays where the past is made to appear more real through sounds and smells but where it is not possible to linger. Others seek to give direct involvement through play and simulation. As we have already stated, the discipline of the nineteenth-century museum, with its cased and catalogued exhibits and educated visitors, has been transformed into a much broader interaction with heritage. However, it is still very much a space of social control and normalization. In each museum a sense of time is inscribed in its internal spatial design: 'the amount of space is decided by the significance and importance of the event, and reciprocally marks the event as significant and important' (Crang, 1993, p.6). However despite the appearance of internal temporal progression, many museums are a normalized series of 'arbitrarily linked spaces' (Delaney, 1991, p.143). In the Amazonian exhibition referred to earlier:

> We enter the Tukanoan longhouse through the men's door and gradually through the shafts of sunlight coming through the thatch we make out the different activities that are going on – the preparation of coca leaf, the grating of manioc. The women's area at the far end is not life-size but modelled to give a perspective of the sunshine out beyond the manioc processing area and the women's door. (Harris and Gow, 1985, p.1)

This was not stated in criticism, merely in description, but what better example could we give of the physical prioritization of the male point of view – those who entered the reconstruction entered as men and saw the women's space as men; no one could see either the men's or the women's space as women would.

The museum experience often merges into the past for sheer fun. Disney's new project for a $650 million American history amusement park near Washington will combine the usual roller coasters and restaurants with depictions of a generic American history – a Civil War fort, an

Indian village, a plantation. Though Disney's Vice-President, Bob Weis, is quoted as saying 'This is not a Polyyanna view of America. We want to make you a Civil War soldier. We want to make you feel what it was to be a slave . . .,' however, Mark Pacala, the park's General Manager was not so sure, saying, 'The idea is to walk out of Disney's America with a smile on your face' (Walker, 1993). It is difficult to see how the two can be reconciled and one would assume that the latter aspiration will prevail.

But not all museums and historical presentations are commercial propositions, local authorities are urged into conservation and preservation as well as the funding of museums, and local-interest groups with charitable status are active in this field too. The museum and heritage site can serve in the project of creating an identity for a town or district, acting as a means of framing the present and giving it depth, as history is ransacked for the names and themes of streets, public houses and commercial centres (Appleby, 1990).

Conservation of the material fabric of the past is obviously selective in the process of renewal; the past can only survive where it is valued by the present or where it is so placed as not to warrant the expenditure of effort which would remove it. Thus we find old urban fabric associated with both the most- and least-valued environments. The means whereby the transition is made from low value 'old' to high value 'history' is through the process of gentrification. An excellent insider view of gentrification is given by Thompson in his book *Rubbish Theory* (1979), where he demonstrates that only people possessing large amounts of cultural capital can rescue goods and buildings from the category 'rubbish'. Such people may well campaign for the districts in which they live to be designated as Conservation Areas, where their own restored dwellings will be complemented by stylistically sympathetic pavings and street furniture and where repairs to houses made in traditional materials will receive public subsidy. It comes to be a matter of financial as well as aesthetic concern to campaign for the historical validity of periods ever closer to the present and for the value of a range of building types. The success of a campaign will, obviously, turn in part upon the imageability of the period and its lifestyle. Conservation areas involve localized gentrification processes which may well involve the search for mythical traditional communities in the past as residents seek out the personalized history of their houses and neighbourhood, as witnessed in the boom of local history societies and genealogical work. Whilst the processes that lead to the wholesale gentrification of areas have become a fiercely contested intellectual terrain (see the debate initiated by Hamnett, 1991/1992; Clark, 1992; Smith, 1992), many smaller enclaves have been gentrified alongside the nostalgic turn towards heritage. But, as Bondi (1991) has shown, the idea of a unified, homogenous group of 'gentrifiers' is both simplistic and misleading. Women are prominent urban gentrifiers, and not just as wives, as they seek to escape suburban entrapment. The irony is that the styles and artefacts they use in their flight from modernity are those of an era in which women had little autonomy or mobility.

Not all gentrification, however, involves the reclamation and refurbishment of the genuinely old. The domestic postmodern frequently takes the form of traditionalism and vernacularism, often whimsically combining several historical periods in a catch-all 'olden times' style (see Fig. 3.1). This has been consciously extended across large areas of urban regeneration like London's Docklands and overflows into new suburban developments seeking to deny the modernist connotations of the suburban by emulating the traditional rural.

We can see the process of gentrification as part of the reterritorialization of the different values that are being invested in heritage, community and tradition, by different individuals and groups in *marginalized* places that remain *central* to geographical imaginations (Shields, 1991). The large-scale transformation and gentrification of London's Docklands still has not completely erased its place-myth as one of working-class, close-knit communities of Eastenders, despite the new marginality associated with *arrivism* and insubstantiality. As Shields (1991, p.256) points out, 'even when the characteristics of a place change so radically that one would expect a change in the place-myth, this does not always take place'. This may well be because of the relative stability of core metaphors and stereotypes in people's memories.

We are conscious that we may seem cynical in our attitude to the way in which the past is inscribed upon place; in part this cynicism is based upon our dislike of the 'faking up' of history involved in its commercialization, the way in which the old is rarely allowed in unless it is cosmetically treated. But this matter of fakery and authenticity is interestingly treated in the context of American museums by Eco (1986) in his seminal paper 'Travels in hyperreality'. With regard to the Getty museum, a copy of a Roman villa he asks the question, 'How do you regain contact with the past?', pointing out that the Roman collectors of fake Greek statues themselves lived in Hellenicized residences. We are also unmoved by the sentimental view of the past as a repository for some lost essential moral community. In Chapter 2 we challenged this romanticized view of tradition as a device for claiming rights of exclusivity and retreat. However, we know that there is an ambivalence in our stance. Our objections to commodified or nationalistic presentations of history should be seen as an attempt to step outside today's structures, not a rejection of the importance of the study of both general histories and genealogies.

In genealogy, selectivity is an important part of the process of constructing an outcome; the perversions of the past upon its past become themselves worthy of our attention. In this context *The Invention of Tradition*, edited by Hobsbawm and Ranger (1983), pioneered the popularization of a recognition that the past is always malleable. In this collection a paper by Trevor-Roper (1983) on the construction of the High-

Fig. 3.1 Dickens Public House, St Katherine's Dock, London

land myth was particularly powerful. It focused on the way in which familiar Scottish tradition with its emphasis on clan tartans, kilts, bagpipes and bardic literature was a product of late eighteenth century, early nineteenth century, predominantly Lowland Scots' desire for a glamorous heritage. The tartans came from English weavers and were assigned to clans only when Sir Walter Scott organized a gathering of clans for the visiting King George IV; the kilts were a work uniform issued by an English iron-smelter employing Highlanders; the poems of Ossian were a blatant fake and so on, in a demolition of everything 'everyone knows' to be Scottish. The elements may not have been traditional, but they have certainly become so. The 'exposure' of the 'real' history behind the 'fakes' took on an added frisson when, not long after publication, Trevor-Roper's expert authentication of 'Hitler's' diaries was discredited after a scientific analysis of their paper and ink and we became aware of the endless chain of simulacra which makes up what we take for the past.

In contemporary society, with its acute awareness of unsettlingly rapid change, there is a reverence for the past and a desire to engage with it. The past may be a 'foreign country' for Lowenthal (1985) but this is a strange metaphor to employ, given that the past is not a thing but a process; is not discontinuous and bounded, and that 'stepping back in time' can be attempted either by staying in the same place or by journeying to places described as being where time has stood still. History can also be regarded as existing without the need for a step backwards at all. This is the view of a movement of French ethno-historians who employ a genealogical perspective to understand present-day places. Sometimes this genealogical work is literal, as in the case of Segalen's *Fifteen Generations of Bretons* (1991), at other times the genealogy is metaphorical, in the sense in which Foucault used the term to develop his idea of a multilinear historical analysis that identified the deviations and accidents within an event. In a process of recovery from reminiscences, festivals, mementos, photographs, stories and proverbs, local practices and artefacts these histories emphasize what persists. A stunning example of this is Zonabend's *The Enduring Memory* (1984), the study of a village in Burgundy. She concludes her book with these wise words:

> The time of the community does not have to explain the present or foretell the future, nor does it vegetate under the weight of the past. Its function is to create a time-span in which the group can work out its own life. There is a stability that is essential, an exceptional individuality in which each group invents its own history and has a memory that is entirely personal and differs fundamentally from the memory of the next social group. In fact, in these societies where forms of sociability magnify differences, this form of Time helps to create the Other. The collective memory conceives the notion of otherness, where possessions of a history that is not shared gives the group its identity. (p.203)

When practising cultural geography one may well run into conflict with historical geographers, whose purpose it is to attempt, like historians, to get as close as they can to the actuality of the past, sometimes for its own sake, more often to explain present-day landscapes or spatial practices. In the task of finding out about the 'real' past, work such as Zonabend's is likely to appear lacking in rigour. But to think this is to miss the point. Cultural geography is concerned with the appearance and meaning of the past, there is then no view which is more accurate or authentic than any other; if people believe in the existence of Robin Hood and his Merrie Men and this belief inscribes itself upon the landscape of tourism or on popular resistance to landlords, then they exist in popular cultural history and geography.

Perhaps we might end this chapter with a proverb, quoted by Zonabend in the preface to her book: 'When memory goes out to pick up dead wood, it brings back the faggot that it likes'.

4

Imaginary places

We hope that by now it is becoming clear that all places are imaginary, in the sense that they cannot exist for us beyond the image we are capable of forming of them in our minds. Said's work *Orientalism* (1978) and Anderson's study of emergent nationality, *Imagined Communities* (1983), were both influential in making people realize that they could not exclude the imaginery from what they liked to think of as reality in geography. But for the moment we want to set this thought aside in order to contemplate those places which are acknowledged to be the self-conscious fabrication of the imagination – fantasy places which one can only go to in the mind, places with no physical presence, places ostensibly conjured out of thin air by creative thought, for we feel that the land of fairy tales needs to be taken as seriously as any other country.

We will find imaginary places in art and literature, folk tales, films and advertising, but we will also find them in the realm of religion and political rhetoric. Invariably, because imagination talks to experience, we will find a correspondence between the fantastic and the mundane and we will find that this reflects back into idealism, morality and normality. Lands, realms and spaces are often constructed in order to defy the physical constraints of experienced materials and social and political structures, but they nearly always lead us back into our own world with its unreflexive imagination. Often they provide us with a cartography, as of Middle Earth or Narnia, even more frequently with a cosmology (such as the Canopean 'Substance-of-we-feeling' or the Narnian longing for Aslan) which is associated with the mystification of a respect for nature and a confusion of conventional categories.

The cultural form loosely referred to as postmodernism (see Box K) has been particularly receptive to a wide spectrum of fantasy. Fantasy always offers a weird mixture of the familiar and the novel. There is a partial recognition of one's own world, but a parallel tracking which distances it from the mundane.

In the relatively simple case of computer games we have the assumption that the game corresponds to a complete world, sealed off from ours

BOX K Postmodernism

Clearly, we are not going to define postmodernism here. We wish only to show some of the debate that this term has been given and differentiate it from postmodernity. In this context we find Cloke *et al.*'s (1992) differentiation of postmodernism into postmodernism as attitude or theory and postmodernism as object useful, though not exhaustive. However, we first must be clear as to what we mean by modernism.

Modernism can most clearly be seen in the 'functional' and 'efficient' architectural designs of Mies, Gropius and Le Corbusier. Driven by a belief in Enlightenment progress, these architects inspired many of the large housing estates built in Britain in the 1950s and 60s. In contrast to the blandness of modernism, postmodernism has been characterized by architects such as Jencks (1987) as a playful pastiche of styles, symbols and images. However, the postmodern city depicted in Raban's *Soft City* (1974) is much more complex than this. It involves a more general recognition of the juxtapositioning of spatial difference in the city, of heterotopias, sites that are in themselves incompatible. This has pointed geographers in the direction of the local scale and an attendance to detail in their analyses. Hence, we have postmodernism as attitude, characterized by Lyotard (1984) as an incredulity towards modernist overarching theories or metanarratives, growing out of postmodernism as object. But a theoretical postmodern standpoint does not just acknowledge that different viewpoints exist in some sort of liberal relativism, it involves a more critical engagement that investigates how difference is actually produced in society through processes of inscription, as well as giving considerable ground to an ethnographic polyvocality (Crang, 1990). It is, thus, not a case of just adding a gender component to existing theories, for example, but of transforming theories by breaking down dichotomies. In this way gender could be incorporated as a fundamental facet of society in such a way as to respect the individual's feelings and desires (Bondi, 1992). However, we must conclude with the observation that in neatly dividing modernist and postmodernist ideas, we are carrying out an extremely modernist tactic (Dear, 1988; Doel, 1992).

and entered as the game commences. A tiny cast of stock characters (often drawn from fantasy films, from the future or from the mediaeval period) works its way through a series of 'levels', each represented by a distinctive environment usually known as a zone or realm – the underwater realm, the ice realm, the urban zone are common examples. To those not addicted, the fantasy seems a thin and unsatisfying one, but to those who are drawn in, the worlds of computer games are compelling, offering people with restricted power the chance to fight off monsters, leap chasms and vanquish end-of-level guardians to get to the end and to win. There may not seem to be much of the world of experience in these games, until one looks at the moral codes contained within them – the equation of power with force; the paucity of female 'characters'; the very act of setting the player up as a hero with a quest, journeying through alien realms. These all reinforce contemporary structures of power with the consequence that, though there are female game players, they are a tiny minority.

The computer console with its platforms and simulations is, of course, only a step away from the military uses of the same technology and although the latter are far from being fantasy in terms of their outcome, they can seem remarkably like it. Baudrillard (1991a, 1991b) argued that because for many military personnel the Gulf War largely took place on computer monitors, and was presented through the media as a computer-game fantasy for the Western public, the war was no longer real. It had entered the realm of fantasy to such an extent that it began to lose any meaning that it had initially and had become hyperreal. At another level, the war was also a playing out of various Western Orientalist fantasies constructed mythologically. We can begin to see that the realm of the imagination can be an extremely powerful tool in society. It is involved in ostensibly creating conflicts out of thin air, but it is also powerful in structuring both dominant and, as we shall see later, transgressive myths.

Much of what we can find in the way of imagined places we can understand as mythological, that curious category of thought which is accepted as unreal and yet true in a transcendental sense. Lévi-Strauss (1962) introduced us to the idea that myths are constructed, through a sort of '*bricolage*', out of totemic elements which are '*bonnes à penser*', a phrase which uneasily translates to mean 'good to think' (not 'think about' or 'think with'). Lévi-Strauss was getting at the way in which myths are self-evidently fantastical but also give full reign to the thinking capability of the societies which generate them, exploring ideas to their point of illogicality, testing the power of opposition by creating crazy oppositional categories, pushing towards the possible-to-think with the impossible (socially or physically) to do. So we find myths granting life to inert matter, giving wings to humans or speech to animals, allowing men to reproduce parthenogenically, having women reproduce with their sons.

Lévi-Strauss' view of myth has always been structuralist and he used mythology to reveal what he considered to be the deep structures of society, but there is no reason why mythology has necessarily to be pressed into service in this way. Indeed if myths are composed of elements which are *bonnes à penser*, they would have to be capable of being used for generating interpretation across a range of possible meanings over time and space and of being pressed into service for a variety of purposes. So we encounter them working in entertainment, the teaching of moral codes, reinforcement of political structures, illustration of passions, the demonstration of aesthetics, the selling of goods and the practice of psychoanalysis. Not only are myths set in fantasy places but they work mytho*logic*ally in themselves (Leach, 1976). But there would be little point in pressing them too hard with the logic of the experienced world; instead we should use them as *bonnes à penser* and see them not only as imagin*ed* but also as imagin*ary*.

The great expansion in exploration which accompanied the early stage of imperialism prompted a genre of Western mythologies which focused on the idea of 'lost worlds' and 'hidden cities', to such an extent that tangible places on the margins of the British Empire began to be endowed with quasi-mystical appeal. The 'lost city' genre serves to divorce civilization from change and modernity and in doing so incorporates technology in nature, rather than opposing the two. Continuity becomes a major theme, with the assumption that there has been a fall from grace in modern times. Nowhere did this apply to more than Tibet, the land walled in by the sacred Himalaya, close to the sky, liminal between great states, dominated by monastic Buddhism. This imaginary Tibet was to be popularized as 'Shangri-La', a benign world of peace and longevity represented in Hilton's novel *Lost Horizon* (1933), which reached mass audiences when it was made into a film.

Bishop's meticulous study *The Myth of Shangri-La* (1989) shows how various outsider groups through different periods of history used the secrecy and enclosure of Tibet for a range of imaginative purposes. It was variously a goal and a threat, it could be used to signify mystery, tyranny or eternity, and could stand for Westerners as archetype of the Orient. The Shangri-La version of this works as a haven of civilization within the snowy wilderness and as a message harmony in a materialist world threatened by war, but it also contains within it the idea of an enchantment which, like a narcotic, can result in the loss of the individual to her/his own culture. Shangri-La is not a pseudonym for Tibet, but it appropriates and builds upon the fantasies which have adhered to the 'real' place; this process in turn serves to mystify the place further, such that it ceases to have any mundane existence for most Westerners, but suffers the indignity of becoming a fairyland, known too well, but never understood.

Shangri-La was more than a distinctly mythical place, however. It was a traveller's story about pilgrimage and liminality, the sacred journey which takes the individual dangerously far from the home structure and then either destroys her in the normlessness or returns her, renewed and wiser. In many other societies around the world we find similar liminal journeys. Meyerhoff's (1975) account of the Huichol Indians is a particularly good example. The Huichol are poor peasant cultivators whose legends say they emerged from a desert land which was previously benign and plentiful. Annually people go by bus to the desert, which they call the land of Wirikuta, to hunt for peyote buttons from which mescaline is derived (they say 'hunt' as if they were animals) and to experience a homecoming to the land they had been forced to leave. They go as pilgrims and throughout the trip their home identities are submerged – they use pilgrim names, insist that there are no divisions of age, gender or status between them. They make their language nonsensical by jumbling or opposing words (so the sun becomes the moon). They fast, but claim that they have been in a land of abundance and have to be dragged out of Wirikuta by the shaman who accompanies them, lest they be tempted to stay for ever in their paradise and abandon all their duties to their relatives and community.

The delights of Wirikuta are imaginary and hallucinatory and they are the experience of a free-flowing anti-structure, which, though it cannot be maintained in the mundane world, does offer an implicit critique of its preoccupations. Turner (1969, 1974, 1978) referred to this anti-structure in pilgrimage and ritual as 'communitas', a spontaneous experience of being at one with others, undivided by rules, categories and logic, and he associated a longing for communitas with a marginal status in the structured society. This marginality might be poverty, powerlessness or despair, but Turner saw its emotional alleviation in a movement to the symbolic margins of the geographical and social world.

In Britain, a group of nomads known as the Dongas tribe has sought to move to such symbolic margins. They look to 'golden ages' and 'lost traditions' in a ecotopian nativism, pushing into the misty past of the European Dark Ages, playing upon Arthurian legend, Celtic myth, American Indian myth and paganism in order to construct a world which is part familiar but also unknown enough to be unpolluted. This has permitted the imaginative reconstruction of a series of environmentally based rituals that work through some of the same themes of liminality and communitas that we saw with the Huichol, despite substantial commodification (see Fig. 4.1). Often these rituals are based upon 'New Age' novels that fantasize about an unpolluted earth, free from patriarchy. Moody (1991) has shown how the fantasy of Celtic mysticism has been particularly appealing to women writers who have found in it sufficient evidence to assume that the point of conversion to Christianity, which is associated with King Arthur, was also the end of legitimate matriliny and equality between the sexes. Celtic Western Europe can then be represented as a place where, in the guise of queens and sorceresses, women may be endowed with power over men.

The valley created by Le Guin in *Always Coming Home* (1988) is a fascinating product of 'New Age' fantasizing, and one which we can see

Fig. 4.1 Avalon, Fawcett Road, Portsmouth

informing many emerging energy, dietary, medical and child-rearing practices. Le Guin's 'future archaelogy' of the Kesh people is set in northern California, but the place is unrecognizable now that it is populated by gentle matrilineal people with a deep spiritual relationship with their environment. In this exercise in 'ecotopia' we find that there is no structural discontinuity between humans, animals and the earth so, in true totemic fashion, a matrilineage contains not only its own domestic animals, who (not which) are relatives, but also such things as water springs or the moon. The category of kinsfolk is prior to the category animal, across which it cuts. Le Guin's story is written as multivocal ethnography, with 'expert' outsider accounts of kinship systems and the like, local 'informants' verbatim accounts, native poetry, pages of music, descriptions of ceremonies and material culture, maps, a glossary and so on. If the jacket of the book did not say that it was a Utopian fantasy it could easily appear on a cultural geography student's reading list. In the valley 'wealth consisted not in things but in the act of giving' (p.117) in evocation of a great many gift-based premarket economies, and the entire book focuses on the possibility of a social order which is non-exploitative and nurturant. It is obviously a highly romanticized version of what used to be called 'tribal' society, however, and the fantasy of a people attuned to nature is one which cynical culturalists like us find a little difficult to be carried along with.

Rather than looking purely to the past, much of New Age literature is about a fusing of ancient, often recreated legends with futuristic, spiritual scenes. Robinson, for example, another ecotopian writer, problematizes a future of space exploration in a moral and political debate which can be seen as a metaphor for Western 'development' of the Third World. In the first book in a trilogy about the colonization of Mars, *Red Mars* (1992) has settlers from earth planning the 'terraforming' of the planet, whilst a group of rebel ecological purists insist that it is immoral to introduce a breathable atmosphere, plants and the rest. In the second book, *Green Mars* (1993) the terraforming has been partially completed and all but the most hardline rebels are fighting a more minimalist position where a thin atmosphere will mean that whilst colonists can live in the hollows without face masks, the high ground (physically and morally) will remain Martian. (The third book, *Blue Mars*, which will focus on bringing oceans to the planet, has not yet been published.) The books become a vehicle not only for moral questioning about the relationship between humans and nature, and situations in which the earth's environment is reproduced scientifically, but they also allow for a wealth of scientific information to be communicated in more or less palatable form.

Ecotopias are, however, only a sub-genre of that most reflexive of the genres of fantasy fabrication, Utopianism. Utopias have a long history, stretching back not only to More's *Utopia* (1516), which coined the term, but further into religious texts about promised lands (whether or not 'flowing with milk and honey'). The Utopian text is always a fairly explicit critique of the world of experience and focuses on the perfectability of humanity (though it is often interesting to look with wonder at some people's idea of perfection). The made-up name of the fictional place, Utopia, draws upon a fusion of the Greek *eutopia*, meaning 'beautiful place', and *outopia*, meaning 'no place'. It acts as a displacement where ideas about perception can be worked through in the form of a narrative. Wilde's often-quoted view of Utopia is worth remembering when configuring cultural geography, for he territorializes it firmly within our academic remit: 'A map of the world that does not include Utopia is not worth even glancing at, for it leaves out the country at which Humanity is always landing. And when Humanity lands there, it looks out, and, seeing a better country, sets sail' (1973, p.34). Utopias are always voyages out of the limits of the present world (Marin, 1991). We can see Utopian writing or film as a form of social, political and cultural experimentation not only with supposedly big issues of structure, but also with the petty details of an everyday life in another realm; an experimentation with both molecular and molar realms.

In the same tone, Ricoeur (1986) contrasts Utopia with idealism, saying that Utopian models are revolutionary, in that they are built upon the idea of a change from the structure of experience, whereas idealism is a means of perfecting that structure through refinement. Where an idealist seeks to improve upon ideas that are already being used to think with, a Utopian does something much more radical and imagines into other structures (potentially, we would suggest, even into anti-structure). Consequently, we can see that enthusiasm for the genre of Utopian creative work often goes in cycles and that there are peaks of interest at times when some sections of society are craving a transformation in their existence, their relationship with others and/or with the physical world; this does not just mean a change in one's lot, but a change in the order of things. The revolutionary power of Utopian fiction is that, having gone to Utopia in one's head, it is a relatively small step to campaign for Utopia to be established in one's own land, with potentially dire consequences for those who are in control under the present system. However, in that Utopia always has to be *somewhere else*, in time or space, there is a degree of safety in shifting the longed-for perfection into the realm of fiction and out of that of social commentary or political treatise. The author of a Utopian text can claim 'only' to have been playing with 'what if . . .' ideas about alternative ways of living when accused of criticizing established ways of being; although, there is often a naive quality to this assertion. The power of subversion contained within fantasy has long been recognized by state governments, which have a tradition of banning Utopian writings, from Campanella's *City of the Sun* (1602), which incurred the wrath of the Inquisition, to Zamyatin's *We* (1924), which displeased the authorities of the Soviet Union (Kumar, 1991).

Practically all Utopian writing at present is oriented towards subaltern groups in 'developed' countries – particularly women and children – for these still cherish the hope that their here and now cannot be the best possible world and that, as Deleuze and Guattari argue, there must be a 'line of flight' out of contemporary problems. Many of the older versions of Utopia written by men seem depressingly 'goody-goody' today with people depicted as joyfully following rational systems of benign law (a bit like a perpetual Scout camp). They often evoke a trust in the personally improving powers of socalism, which in its simpler forms no longer holds out a convincing promise of a harmonious future; in short, they seem dated – we have 'set sail' from them. Insofar as the late twentieth century still constructs Utopias, they are predominantly feminist (even if this implies dystopian pessimism for some men)!

An intriguing early feminist Utopia was constructed by Gilman in her book *Herland* (1979). Written in 1915 as a serial, this story needs to be read in the light of her classic novella *The Yellow Wallpaper* (1892, 1985). Where the female protagonist in *The Yellow Wallpaper* is trapped in space and structure, the women of *Herland* dwell in an expansive state of freedom in a realm without men and where reproduction is parthenogenic. Motherhood, not power, is the central value of the society, but this is a motherhood which does not operate in opposition to paternity. Without this opposition women exist not as others but as complete humans so that when male sociologists mount an expedition into their territory they find that women are both physically and behaviourally different from those they have previously experienced. The feminine luxury items they have brought to win friends amongst the residents fail to impress these tall, strong, autonomous, beautiful women. They are surprised to find that without the influence of men, age differences between women become unimportant and notions of beauty are no longer associated with youth. The economy of *Herland* is created out of the maternal notion of nurturance and the creation of abundance. But it would not be appropriate to refer to *Herland* as matriarchal, because the whole idea of the need for rule is challenged; it is assumed that a realm based on love can survive without a structure.

The book acts as a vehicle not for working out alternative structures (the initial premise of a reproducing world of women was only imaginable, not realizable), but for speculating about a world where women's thinking was not generated dialogically with men's. It offers up intriguing thoughts about the nature of love and wealth as the women of *Herland* try to explain their values and practises to the explorers and it presents us with a terrible feeling of betrayal when the men foist their possessive and exclusive notions of love on the women. When the men explain to them about home, privacy, possession, living together in couples, the women cannot understand why any of this should be thought good or how it can enhance their mothering.

Gilman's Utopia would not appeal to everyone – she dreamt of a reproductive system without sex, when many women around her desired sexual freedom she offered a sacred motherhood. However, the fantasy, in its resonance with Gaian notions of Mother Earth, protection and fruitfulness, still has a certain power in the late twentieth century eco-feminism and the book is enjoying an unexpected revival in popularity. Piercy's *Woman on the Edge of Time* (1979) and Russ's *The Female Man* (1975) offer much more recent examples of the Utopia constituted in the feminine, challenging our abilities to think gender categories without opposition and to think structures without boundaries (see Box L).

BOX L Fantasy

Writing of the fantasy depicted in the genre of women's film of the 1930s and 40s, Doane (1987a, pp.182–3) argues that.

> it is only through a disengagement of women from the roles and gestures of naturalized femininity that traditional ways of conceptualising sexual difference can be overthrown. Mimicry as a textual strategy makes it possible for the female spectator to understand that recognition is buttressed by misrecognition. From this perspective, fantasy becomes a crucial site of intervention and what is at issue is the woman's ability to map herself in the terrain of fantasy. . . . The fascination which the women's films still exert on us can be taken up and activated in the realm of fantasy rather than melodrama – particularly if the fantasy is perceived as a space for work on and against the familiar tropes of femininity. Because everything depends, of course, on how one sees oneself. And it is now possible to look elsewhere.

The fantasies Doane was considering did not involve the construction of displaced and imagined worlds, they looked into the pathos and romance of a familiar world of dreaming in the here and now, and her interpretation of the role of fantasy is ever present in the genre of science fiction represented by Atwood (1987), Lessing (1979) or Piercy (1979).

Building upon the style of the Utopian fiction is the dark side of the genre, known as 'dystopian', playing upon the idea of the future having gone awry, where Utopian-inspired planning and progress has got out of hand. If Utopian fiction is revolutionary and experimental, its dystopian twin is often conservative and cautionary, using imagination to reinforce existing structure. Utopian invention was popular in the optimistic days of nineteenth-century modernization and was built upon the dream that new technology could free human beings from both poverty and drudgery. However, the twentieth century has been more persuaded by dystopian pessimism in the wake of two world wars, totalitarianism, the nuclear threat, and the ecological crisis.

One of the most popular dystopian constructions of this century was Huxley's *Brave New World* (1932). The underlying theme of this novel was that the planned Utopia, with its obtrusive rule system and reliance on the power of technology, would end in the dehumanization of those who inhabit it. In the world presented by Huxley most work has been abolished by processes of automation and that which remains is done by people specially bred for their brutalized role so that they will not question their lot. Sexual pleasure has been separated from both procreation and notions of a long-lasting partnership, and the family has collapsed as the state provides for both the needs and the desires of its people. The whole society rests upon a hierarchy of genetic endowment and scientific rearing which grades people from Alpha Plus to Epsilon Minus/Semi-Moron in an ascriptive caste structure. The story is a fairly explicit critique of the planned economy and rational society of the Soviet Union at a time when it might have seemed enticing to a large number of British men. However, in equating planning with loss of virility and family planning with the demise of the family, *Brave New World* also glorifies an escape into nature (and natural fertility) out of artificiality. The character of the Savage, born 'naturally' by a woman rather than out of a bottle and conceived out of love rather than out of science, is both a product of nature and high culture (he can recite Shakespeare). The novel plays upon patrician ideas about the English country gentlemen (a blend of natural superiority, based upon cultivated breeding) in opposition to the self-made middle class and the urban, industrial proletariat with its 'mass' culture. The book successfully captured a great many establishment fears in one fell swoop, not least the fear of the power of women released from parturition and nurturance.

Orwell's *Nineteen Eighty-Four* (1948) relied on distancing in time rather than in place and has rather doubled back on itself now that the date has been passed. This dystopia relies on technology for its fantastical elements, pushed forward into a condition of global totalitarianism. The menace in this imagined world is that of absolute technical surveillance under a mendacious bureaucratic system; the horror of a propagandizing television which cannot be switched off, and which can see as well as be seen. Individuality is systematically erased through a combination of double-think conditioning, a ban on writing by hand and censorship of art and history. Much of the power of *Nineteen Eighty-Four* was derived from the sheer ordinariness of the physical environment in which it was set; the drabness of postwar London was captured perfectly and then pushed a little further to make a fitting home for the depersonalized, inhumane society Orwell feared. Like all dystopian creations the book was highly political; the assumption was that if we were not very careful, the future could be a great deal worse than the present.

Contemporary dystopias are most commonly portrayed in film, that space of visual desire that works ostensibly because it holds a tension between the specular space – everything we see on the screen, and the blind, imaginative, space that is everything happening under the surface of the screen (Telotte, 1990). We can see both *Bladerunner* and *Terminator* playing upon the fear of a loss of humanity in an increasingly technologically oriented world. The power of the cyborg (cybernetic organism) as a contemporary, postmodern totem is that it is simultaneously more *and* less than human – an inhuman 'machine' made from programmable genetic material. It makes us ponder the limits of our own humanity in a world where nature is subjected to artificial control.

There is no doubting the power of *Bladerunner* as a key dystopia for our times. The story of the hunting of escaped cyborgs, called replicants in this case, is set in an unrecognizable Los Angeles, where familiar multinational firms are advertised on orbiting screens alongside propaganda for off-world colonies and where street activity takes place in a global patois leaning heavily on South-East Asia. The world of *Bladerunner* is one of multiple simulations and simulacra where the notion of reality becomes almost irrelevant as all of our normal tests of reality become palpably unusable – sight, touch, memory, all can be simulated (see Box M). In the final sequence of the *Director's Cut* version we are forced to abandon our distinction between humans and cyborgs; watching the cyborg woman, supposedly granted a 'real' life as she elopes with the supposedly human Bladerunner. The dystopia of *Bladerunner* has its roots in the great ecological questions which seek to understand status of humans in nature as well as the relationship between humans and nature. It follows the familiar cautionary form of the present running away with itself, and reflecting back 'normal' reproduction and patriarchal relationships as the only salvations from a dehumanizing future.

Reproduction of humans or humanoids through biotechnology in one form or another is a recurrent dystopian theme. It is inevitably associated with the absence of the family as a basic unit of social organization and with emotionally distant but all-seeing controllers. Perhaps we could suggest that dystopia is characteristically a place where man (no, that was not a slip) is denied the luxury of Oedipal desire, where the father becomes all powerful and the mother never existed, even as an idea. It should not be too surprising that similar themes can be worked through in a different way. Feminist dystopias, like Atwood's *The Handmaid's*

BOX M Simulacra

Baudrillard (1990) has argued that we are entering an era of infinite spatial simulacra. By this he means not a simulation which is merely a copy of a real thing, a photo is a copy of a landscape, but a copy of a copy, *ad infinitum*. With increasing mediatization, the power of the image has accelerated through time-space compression and commodification to the point where we no longer know how to distinguish between various copies and originals. The very notion of the original has become obsolete, as we no longer know whether it may be just a fake, like the replicants in the film *Bladerunner*. The only way that we can cope with this hyperreality, argues Baudrillard, is to take it literally and enjoy it.

At first sight the idea of a simulacrum may seem close to Debord's concept of the spectacle, which he developed in his book *The Society of the Spectacle* (1970). Debord argued that in late capitalism, a confusion between images or representations in the widest sense and real things was a necessary condition for its success. However, unlike Baudrillard, Debord argued that the integrated spectacle of capitalism was not a seamless simulacrum; he retained the possibility of political intervention when he argued that the spectacle 'is not a collection of images, but a social relation among people mediated by images' (p.4). Thus new social movements, such as punk music, could shock and subvert the capitalist spectacle, even though they would probably be recuperated back into capitalism through the proliferation of kitsch commodities (Plant, 1992).

Tale (1986) are built out of a heightening of conventional sexual roles, such that a woman becomes a slave to the gratification of excessive masculine desires for offspring, or submission in structured form.

In Atwood's story the women of Gilead have lost all their autonomy to men – a men's revolution has deactivated all of the bank accounts and credit cards marked 'F', replaced women's jobs with 'tasks'; relegating them to dependence upon masters and dividing them into classes of femininity, for example, 'wife', 'handmaid', 'Martha' (domestic) or 'aunt' (teacher of female roles). The world, unlike the one we live in, is *entirely*, not merely predominantly, constructed for the benefit of older men. Patriarchy is pushed to awful limits; mothers are relegated to a fragment in a system of division of labour, and biological surrogacy becomes an accepted necessity. Handmaids are kept for reproductive purposes, to conceive and bear the children of the many infertile wives and to do the shopping for the households of their male 'commanders'. They have no identity of their own, being known teckonymically through their commander (so Offred is Commander Fred's handmaid). Relations between women, whether of the same or different ranks, become centred on envy and competition and the whole society is predicated upon an extreme form of rational economic scarcity. Abundance and luxury have been pared down in an economic puritanism which assumes that 'A thing is valued ... only if it is rare and hard to get' (p.123). The men had initially risen up against what they saw as a growing indulgence on the part of women, manifested in a declining birth rate, which history shows to have been caused by increasing pollution of the world, not, as the men of Gilead think, by women's right to choose.

As a dystopia Gilead works by pushing phallocentric values of authority, patriarchy, law, scarcity and purity as far as they will go. There are quite overt parallels with the conventional Westerner's view of Islamic society – women veiled, their limbs covered, fertile puritanism, strong household heads and patrilineal systems. Women have come to believe that the structure in which they live is moral and neither they nor the men have a language or repertoire of symbols to conjure up a critique of their society; even more pathetically, they have stifled imagination with excessive logic and overstructuring. Gilead is not just a land of rules, it is also a land of boundaries, physical confinement and surveillance. Handmaids have their own rooms and they are well provided for (it is essential that they are healthy enough to conceive) but there is no sharing, no commensality and certainly no love. They go on shopping trips in pairs, but they are co-chaperones, not friends – they exist to confine each other, spatially as well as socially. Daily they must approach the wall that not only separates the city from its hinterland and serves as a boundary which is impossible to cross, but which is also a place for displaying the bodies of those executed for transgressing the moral boundaries of the society. Whatever their rank, the women of Gilead are confined – wives, trapped in shallowness and genteel idleness, cannot go anywhere beyond their own homes, other than to visit the sick or those giving birth; Marthas remain working in the house; handmaids go out but, like schoolgirls, only under licence and wearing their distinctive uniforms.

Gilead exists in a state of permanent war, experienced (like Baudrillard's Gulf War) only on television. It is not suggested in the novel, but one suspects that it is not actually happening other than in mediatized form in order to unify the people of Gilead against outside enemies and to act as a rationale for the shortages and rationing of goods. Daily life in Gilead is sparse and pared down, sensorially deprived and sanctimonious; the nation encapsulates itself, claiming that previous eras and other countries are decadent. There is a terrible fear of the consequences of earlier pollution:

> The air got too full, once, of chemicals, rays, radiation, the water swarmed with toxic molecules, all of that takes years to clean up, and meanwhile they creep into your body, camp out in your fatty cells. Who knows your very flesh may be polluted, dirty as an oily beach, sure death to shore birds and unborn babies. Maybe a vulture would die of eating you. Maybe you light up in the dark. (p.122)

There is an obsession with purity and control, underpinned by a morbid fascination with disorder, the disgusting and immoral – Gilead fights a constant battle against 'dirt' in Douglas's sense of matter out of place. It is a society which has lost all sense of community in exchange for a rigid system of law and *order*. Gilead is a land which we ought to think carefully about, because in many ways it is too close to reality for many women.

Dystopia, as we have seen, is frequently associated with an excess of structure and technology. Utopia, by contrast, is depicted in its current form as rhythmic, rather than structural, organic rather than technological, expansive rather than conservative, generous rather than calculating.

Although this chapter wanders through the realms of the fantastic, rather than the concrete, we hope that it will become apparent that we see it as pivotal in our argument about the structuring and unstructuring of worlds of desire and realms of power. Although imagination is rarely entirely free from the same structuring processes that govern logic, it is in imagination and artistic creation that it becomes possible to think the unthinkable, to challenge codes and to deconstruct categories.

5

The world as a place: a global culture?

Throughout this book we are very conscious that the world in the last years of the twentieth century is a very different place from that which existed in the late nineteenth century. But it is now time to ask ourselves whether the world *is* a place of any meaningful sense. We cannot find any references to people talking about, for example, 'making the world a better *place*' in the nineteenth century, and it certainly would have seemed an illogical construction for the high point of imperialism when 'places' were still being 'discovered' and colonized. With Mackinder's 'heartland' theory we get the beginnings of a new construction of the world as a functional geopolitical whole, but not yet a place.

To talk of the world as a place is to imply that there is a commonality. The whole idea of a global economy and global politics makes an assumption that there is, to some extent, a global culture – that certain meanings can be understood worldwide. But, as Cosgrove (1991) points out, this is a reflexive thought and the image of the globe can be used to promote a sense of unity and accessibility for commercial and political purposes. He urges us to think about the way in which the terms 'world', 'earth' and 'globe' are used sometimes interchangeably, sometimes differentially – 'world' having the capacity to represent orders within the whole (as when people talk of the 'Third World' or, very differently, the 'academic world'); 'earth' having an ability to seem devoid of encultured human beings, a physical thing; but 'globe' is invariably encultured and whole, though it is symbolized by reference to the earth. We need to question not only whether there is global culture, but also whether such a concept can include the imagination and the practice of everyone, or can be conceived of by all people. We need also to ask ourselves how and why the issue of global culture emerged when it did, why it became interesting to contemplate the overlaying, mixing, fusion, hybridization, or alloying of local cultures.

Most of us were encouraged at sometime to think about the way in which 'our' culture was in fact not a homemade thing at all, but draws upon ingredients from 'other' places which we regard as foreign. Many elementary geography books do this, presumably to encourage children to think about the wonders of diffusion, for in them the account is rarely politicized (see Box N).

BOX N Cultural diffusion

Our solid American citizen awakens in a bed built on a pattern that originated in the Near East but that was modified in Northern Europe before it was transmitted to America. He throws back covers made from cotton, domesticated in India, or linen, domesticated in the Near East, or silk, the use of which was discovered in China. All of these materials have been spun and woven by processes invented in the Near East. He slips into his moccasins, invented by the Indians of the Eastern woodlands, and goes to the bathroom, whose fixtures are a mixture of European and American inventions, both of recent date. He takes off his pajamas [sic.], a garment invented in India, and washes with soap, invented by the ancient Gauls. He then shaves – a masochistic rite that seems to have been derived from either Sumer or ancient Egypt.

... On his way to breakfast he stops to buy a paper, paying for it with coins, an ancient Lydian invention. At the restaurant, a whole new series of borrowed elements confronts him. His plate is made of a form of pottery invented in China. His knife is of steel, an alloy first made in Southern India; his fork, a medieval Italian invention; and his spoon, a derivative of a Roman original.

... When our friend has finished eating ... he reads the news of the day, imprinted in characters invented by the ancient Semites upon a material invented in China by a process invented in Germany. As he absorbs the accounts of foreign trouble, he will, if he is a good, conservative citizen, thank a Hebrew deity in an Indo-European language that he is 100 per cent American (Linton, 1936, quoted in Jordan and Rowntree, 1982, p.14).

We hope that you find this sort of trite example of long-standing globalism as embarrassing as we do. What is fascinating is that it was still being quoted with approval in a mainstream American cultural geography textbook as recently as 1982.

More recently, Harvey has urged us to think about where our breakfast comes from and Williams (1983) has explored the foreign constitution of

Britishness, but these writers are demonstrating the position of Western patterns of consumption and production within a global network of exploitative relations. However, global culture, if it exists, is not a simple matter of imports and exports – tea may 'come from' China, but we know that it means something completely different in each of its Indian, British, American and French settings and we would venture that in all of these four a shifting concept of 'Englishness' often has something to do with it. Once again the Foucauldian metaphor of genealogy becomes useful here, always remembering that a genealogy is not simply a line of descent but also contains unequal collateral kin, who contribute as much to one's identity as do one's ancestors.

The very imagining of a global culture, in the singular, assumes the working of two very different processes, but slithers, dishonestly, between them. On the one hand is the legitimization of hegemony, globalization conceived of as the export of 'superior' cultural traits from 'advanced' countries and their worldwide adoption; on the other is a hydridization and emergence, through dialogue, of new universal cultural practice. Anyone who has ever travelled between major world cities cannot help but have noticed that many things seem familiar whichever country one is in – airports, offices and international hotels may have local variations but they are not radically different. Though women may be wearing a local costume, it is likely that men employed in professional jobs will be wearing a variation of the conventional 'business suit'; relatively affluent young people will be wearing jeans, preferably those made by an international company; rock music is ubiquitous, so is Coca Cola; the same TV shows are dubbed into a host of languages and shown internationally, whilst computer games work across a superficial global/galactic symbolic realm which seems comprehensible everywhere (see Fig. 5.1). It is tempting to see all this as evidence of the 'Westernisation' 'Americanisation' and even 'modernization' of the world and to see those terms as synonymous. We might then ask whether it is not realistic to assume that there is a global culture, particularly for the young. But we should also begin to think about cultural flows in the other direction. One is unsurprised by Indian, Mexican or Chinese restaurants in European and American cities (though a Mexican restaurant in Delhi would be less expected, as it would also be in a Provençal village), or by boutiques in Europe or America selling exotic craft goods. Parisian high fashion periodically draws upon different 'ethnic' influences and these then influence the street fashions of much of the world.

We have to confess to being rather more unsure about how to interpret all this than some of our fellow geographers, particularly those like Peet (1991), whose materialist base is unshakable. Our lack of confidence emerges from our inability (with Foucault, 1984) to see any single thing as *fundamental* in the explanation of anything else – we cannot then fall back upon a straightforward assumption that globalization of capital

Fig. 5.1 Satellite dish on the roof of the City Palace, Jaipur, India

results in the globalization of culture. We do believe that the two are related, however, and that time-space compression and the new international division of labour have important cultural, as well as economic and political, implications.

If we think back to the high point of imperialism, ideas about internationalization of commodities and practices were relatively straightforward and could be theorized according to a straightforward economics of scarcity. The account of the un-American content of an American life makes sense in the light of America's hegemonic role, extracting surplus value from a hierarchy of dependent countries and drawing upon a history of trade and diffusion. We can see the wealthy nations, and the wealthy sections of all nations, as drawing upon the goods of the whole world, as centres of wealth become the centres of flows of goods as diverse as vegetables, artworks and diamonds. The flow of information, however, goes in the opposite direction, with the major capitalist centres acting as distribution points whether of news, data, literature, films and TV programmes, religion or academic texts. We see the rest of the world being manoeuvred into unequal partnerships which implied not only exploitative economic relations but also necessitated the acquisition of new languages, religions, values and lifestyles. This is usually explained in terms of dependency theory or world systems theory.

Local cultures undoubtedly came under threat as élites learned that it was advantageous to cosmopolitanize. Some of the best examples of this can be seen within the context of British colonialism (particularly in India, the so-called 'jewel in the crown' of the British Empire), where local aristocrats were not only admitted into an international social set based around conspicuous consumption and exclusive hospitality, but were also encouraged to educate their children at British public schools and ancient universities. For the middle class, a local experience of a British education system focusing upon the Cambridge School Certificate and the London External Degree served to homogenize the formal knowledge of the Empire. Education came to mean British education and people who had only local knowledge came to be seen as 'uneducated'. The French policy of assimilation, where the acquisition of a high enough degree of French education could result in the granting of metropolitan French nationality, was an even more overt example of the same process of cultural imperialism.

The role of missionaries in spreading not only the Christian religion but accompanying notions of deference, family and individuality is well-known, but we should recognize that this was by no means a passive process and there are ample examples of the subversion of the metropolitan culture in colonial and neo-colonialist settings. The separatist churches and millenarian religions which broke away from missionary orders throughout the colonized world and the ethnic revivals of language, cuisine, art and dress (sometimes in a nativistic form, sometimes as local innovation), are obvious examples. The classic study in this field is Worsely's *The Trumpet Shall Sound* (1968), which focuses upon the Melanesian Cargo Cults, but a stunning Foucauldian account of the history and present resistance and accommodation of South African Churches of Zion is given by Comaroff (1985). Lann's *Guns and Rain* (1985) shows how spirit mediums and the cults surrounding them became a focus for the freedom fighters during the war of independence which brought the new nation of Zimbabwe into existence: local cults took on a global significance. These various movements can be seen as cultural responses to hegemonic forces, offering a valuable rallying point for political protest and independence movements as well as a means of thinking through and experiencing oppositional identities. However, these oppositional identities are not themselves static. So for example, there is a complex relationship between the Rastafarian religion (which looks back to Africa, but does not emerge intact from it), reggae music (with its African rhythms but Caribbean content), urban poverty and Jamaican nationalism. This was then employed politically and communicatively by young black people in Britain; became a vehicle for more generalized anti-establishment protest by young whites; and was commercialized and re-exported in the 1970s. It is clear that the process of cultural 'diffusion' is neither simple in meaning nor unidirectional in flow.

Since culture is a repertoire of symbols, beliefs and practices employed communicatively by groups of people, we should not be in the least surprised to find that, like race, its artificial construction evaporates into thin air as soon as one tries to capture it within defined boundaries. For us it is axiomatic that not only is culture a process and not a thing but, again like race, it is a process which is often treated *as if* it were a thing. As soon as one starts to talk about places of origin and diffusions, with implications about original appropriateness and less authentic 'borrowing' or 'theft', or enforced conferment, one misses the dynamism of culture and begins to think of it as if it were the content of boxes placed alongside one another. Thinking this way, one conceives of an interaction which is no more profound than mixing up different beads in a shop display, all ultimately capable of being sorted back into their 'proper' place. So long as one structured one's thinking about the world in terms of more or less autonomous nation-states, this way of conceiving of culture was fairly satisfactory. Culture *conceived of as a thing* was a means of identifying nations as things (i.e. states) rather than as political aspirations and we find the emergence of modern states and state cultures at the same time, notably in the eighteenth and nineteenth centuries in the case of Europe and for colonized countries, as they emerged from colonial status. It should be apparent that contemporary configurations of culture and power do not easily lend themselves to this reification, though cultural identities will be grasped at and appear tangible at particular moments in the configuration of human interaction, often with terrifying results, as manifested in the term 'ethnic cleansing'.

In answer to the question 'is there a global culture?', we would have to answer 'no'; not just because we believe, like Peet (1991), that it is currently in a state of incompleteness but is on its way, nor, like Hannertz (1990), because we believe that there are simply global cultures, in the plural, but mainly because we are disinclined to think of cultures as corporate entities. What we need to do is to think of the ways in which people with different perceptions and experiences conceive of being in the world, how globalism and localism are configured in contemporary conditions. This configuration means that there will be geographical differences in the assemblage and that these will emerge both from history and from present conditions. People throughout the world make cultural responses to their experience of globalism, working it through in ways which are more or less intellectually satisfying. This means that a common process can manifest itself in different ways, but also that superficially similar traits may have locally different meanings. For some it might seem that these differences of view matter very little, but in theoretical terms it is quite crucial whether one thinks of culture as a thing, an ideology, a response or a random set of practices.

There are many pathways through the observation that the places and people of the world seem increasingly to share cultural practices. Some explanations are predicated upon technological innovation (an assumed 'shrinking' of the world through transport links or electronic media); some upon economic forces; and others upon political strategies and tactics. We see each of these processes as a series of cultural flows, each deeper in some places than in others, with uncertain limits, sometimes eroding what they flow across, often carrying sediment within them. Certainly technology, finance and power are implicated in these flows, but so are comics and video games, classical, rock, and pop music, soap operas and horror movies, tastes in clothes, perfume, architecture, interior design, flower arranging and food. The conceptualization of speed, ultimately to the point of apparent instantaneity, becomes crucial to understanding the unevenness of the flow. As Lash and Urry (1994) point out, the urban system made up of London–New York–Tokyo is a financial interconnection not competition. It becomes feasible to think of the three as the core of a global urban system of concentrated producer services in a geography of speed, not surface distance. For financial purposes they constitute one place, and in doing so they detonate common-sense notions of place and accompanying ideas of being within it. To some extent the totemic youth culture identified with this core is common; children in Europe and America are acutely aware of Japanese hardware and software and popularized representations of Japanese martial arts; often young business people self-consciously aim at a hybrid culture, whilst the middle-aged and elderly often use only that Japanese technology which fits in with older Western forms and are likely to draw upon residual racist categories in their rejection of Japan (Morley and Robbins, 1992). For Lash and Urry (1994), what characterizes this new transglobal core is *reflexive* accumulation, the recognition that culture and economy are not so much linked causally as that there is an aestheticization or enculturation of economic life, particularly for a certain category of service-sector employees. Clearly this idea builds upon but also takes us beyond Bourdieu's (1984) idea of cultural capital and we can begin to see processes of inclusion and exclusion from differently layered assemblages of cultural-economic practice forming a new global hierarchy.

Arjun Appadurai (1990) seeks to impose some order upon the thinking through of the differing depths and speeds of global flows in his conceptualization of interrelated worldwide 'landscapes', constituted through what he refers to as 'ethnoscapes', 'mediascapes', 'technoscapes', 'finanscapes' and 'ideoscapes' – the 'scape' suffix implies a 'looking-from-the-point-of-view-of', rather than a 'structuring-in-the-light-of'. Though his neologisms may be ugly, they are useful as a device for separating out elements of culture for the purpose of observation without then willing any prioritization upon them. His personal view as an Indian writing from within the American academy also moves us away from a structuring of the world on to an occidental centre, as he seeks to understand the complex relationship between people, technology, ideas and wealth, through emphasizing micro-causalities and backwashes as well as the grand structures of investment.

Appadurai's *ethnoscapes* are understood as flows of people – people who travel under many different labels – immigrants, exiles, tourists, expatriates, businessmen and women, refugees, performers, academics, guest-workers – but who have all to come to terms with experiences different from those of home and 'the realities of having to move or the fantasies of wanting to move' (p.297), at whatever scale. A *technoscape* is a 'global configuration ... ever fluid, of technology' (p.297), which he sees as 'driven not by the obvious economies of scale, of political control, or of market technology, but of increasingly complex relationships between money flows, political possibilities and the availability of both low and highly-skilled labour' (p.298). *Finanscapes* are constituted out of the rapid flows of global capital: 'But the critical point is that the global relationship between ethnoscapes, technoscapes and finanscapes is deeply disjunctive and profoundly unpredicatable, since each of these landscapes is subject to its own constraints and incentives ... at the same time as each acts as a constraint and a parameter for movements of the other' (p.298). These disjunctures are what become particularly interesting in cultural terms for it is in them that we find the *mediascapes* – distributions of electronic and other mediated information – and *ideoscapes*, which may be closely intertwined with the images of mediascapes, but which relate to the distributions of ideology. Media and ideology seek to smooth out the disjunctures in flows of people, technology and finance by constituting imagined worlds and looking from the perspective of these. The more one thinks through Appadurai's simple schema and the dialogue between the different views, the more exciting the notion of *difference* becomes in terms of cultural practice. The world can be conceived by some *as if* it is a cultural whole, but the experience of that wholeness varies, for both individuals and for places. People only *possess* a culture insofar as they work with it.

Appadurai gives the example of Indian migrants, distinguishing the unskilled labourers who go to Dubai from the software engineers who go to the United States – both dealing in different ways in different locations with different experiences of Indianness and outsiderness, both potentially importing and exporting different knowledge and contacts into and out of different Indias. Their migration can be interpreted as part of a redistribution of capital and of particular technologies. But their personal, media and ideological connections with India mean that it is constituted as 'home' and it is likely that they will invest money earned abroad there, with implications for the global flow of capital, for the Indian national economy, for local economies and for the appearance of Indian towns and villages. Some of this investment will be in buildings, whether as homes or commercial premises, and construction work will in turn result in new internal and external flows of people, technology and finance. There will be Western architects and construction engineers and ex-peasant labourers from rural areas working, with very different experiences and expectations, on the same projects in the same Indian cities, spending and investing their incomes in very different contexts.

The migrants constitute a market abroad for the products of the Indian media and other cultural signifiers like clothing and food, and the nostalgia for 'home' constructs a particular form of Indianness. This may then return, reflexively, to India as a new form of sophistication, sometimes feeding into cosmopolitan culture via the 'host' countries. We may see this as similar to the way in which a curious brand of 'Englishness' was created by colonials in India during the Raj and was then was not just incorporated into Indian élite culture but was imported into Britain as part of a middle-class culture of exclusion from an increasingly affluent working class. The Anglo-Indian motifs then took on a disembodied quality of élitist nostalgia which still sells internationally as fiction, film and as material product. But the nurturance of identity away from home is not all a matter of nostalgia and lifestyle, it can have very real political significance. Appadurai cites the example of the invention of 'Khalistan' as a Sikh homeland by displaced Sikhs in Britain, the United States and Canada and the implications of this for Indian politics and security.

Appadurai sees the return of migrants as making changes in Indian relationships, not simply because they bear with them imported cultural traits, whether material or immaterial, but also because they bring with them the experience of a particular way of being foreign. So, for example, the men who had been workers in the Persian Gulf had experienced isolation from women which often resulted in a demand there for various kinds of pornography, opportunistically supplied from India and rapidly finding a market. Appadurai sees this as changing local gender relationships, not only men's views of women back at home but some women's views of themselves through the commercial opportunities available to them in Indian cities.

Clearly, the process of international migration is not uniformly experienced even at the same site, neither is it equally represented in global flows of information. Brah (1991) has compared the plight of women from Western countries and Asia stranded in Iraq and Kuwait during the Gulf crisis. She reported that Asian women were likely to be lowly paid employees, losing their incomes and their ability to send remittances home, where they might well have been the sole source of income for an extended family; they left with nothing, dependent upon aid. Western women, on the other hand, were predominantly the wives of highly paid men; they too were to experience a radical decline in standard of living, but not destitution. Whilst awaiting evacuation, the Asians found themselves in tented encampments, the Westerners in international hotels; the latter were interviewed regularly for international news broadcasts, the former were muted and invisible.

There have been various ways in which such differences have been theorized, some more sophisticated than others. Hannertz (1990) draws upon Merton's (1957) ideas about 'locals' in opposition to 'cosmopolitans', but whereas Merton had used the distinction to understand people with different views within a community, Hannertz employs it to think about different conceptions of the global and local continuum. Hannertz sees a local as not necessarily someone who always stays at home, but someone who structures the world according to local knowledge based around familiar people and things. When a local travels he or she seeks out the familiar and the ideal foreign location is one which can be seen as 'home plus' (plus beaches, business deals, sunshine, exotic culture traits, etc.). It is this 'home plus' requirement which is catered for in 'international' hotels with 'international' cuisine, where the staff speak a range of languages; it is also the basis of the sort of travel guide which aims at finding the traveller the least frightening experiences. In contrast, the cosmopolitan is someone who travels, adapting to the conditions of the country he or she is in and treating the difference as something to be savoured. It is the style of travel of a person who prefers not to be thought of as a tourist and who thinks of travel as a form of cultivation of sensibilities. But, says Hannertz, such a person does not fully engage with the foreign place, rather raids it for experiences which gain in currency on return home. He sees the cosmopolitan as a source of movement of cultural practices between localities and someone whose notion of sophistication rests upon knowledge of the world, but he realizes that the cosmopolitan relies upon the foreign local as a source of experience and the home local as a foil. Thus cosmopolitanism is constructive and conservative of localism. Hannertz then concludes that there are many global cultures based around the internationalization of localisms by both globals and cosmopolitans and there is also a kind of corporately based 'third culture' of internationalism.

In contrast to Appadurai, Hannertz's view of 'the rest of the world' is a firmly male, Western one and is devoid of gender sensitivity. Throughout he refers to the cosmopolitan as 'he' and he talks of cosmopolitanism as a kind of 'mastery'. Hannertz's cosmopolitanism is clearly not as subtle as the hybridity of Bhabha (1988) or as sensitive as the 'transculturation' of Pratt (1993), both of which take account of conflict in the creation of cultures of contact.

Massey's (1993) concept of *power geometry* also attempts to explore the ways in which, for people with different access to power, different notions of what constitutes global and local exist. She emphasizes that scale is, as always, important when thinking about the globalization of culture, for the very concept of the global rests upon ideas about differentiation from a range of locals. Massey describes the area of London where she lives and remarks upon the evidence of a large Irish community in the Irish newspapers on display, the IRA graffiti and the sense of *connection* between Kilburn and Ireland. The shop selling saris is witness to a localization of Indian immigration. She also thinks about the global and the local not as culture traits but as experiences of physical connection:

> Overhead there's always at least one aeroplane – we seem to be on a flight path to Heathrow and by the time they're over Kilburn you can see them clearly enough to discern the airline and wonder as you struggle with your shopping where they are coming from. Below, the reason the traffic is snarled up (another odd effect of time-space compression!) is in part because this is one of the main entrances to and escape routes from London, the road to Staples Corner and the beginning of the M1 to the north. These are just the beginnings of a sketch from immediate impressions but a proper analysis could be done of the links between Kilburn and the world. And so it could for any place. (p.65)

It is not just the wonderousness of connection in itself that fascinates Massey but the way in which different people have different access to that connection for different reasons. So we can appreciate that, for some, the plane overhead is local noise pollution; for others, it is the way home; and for others still, a business requirement or holiday opportunity. The experience of time-space compression, signified by the aeroplane, is clearly not the same for everyone, not even those who find themselves in the same location. But transport is today the slowest form of communication in an era of instantaneous global telecommunications.

Flows of information have changed in a great many respects since the earliest days of transculturation. The cost in money or time of superior communication is such that the map of flows of information have become massively distorted. Gould (1991) depicts this as a 'backcloth' of some infinitely stretchable and scrunchable fabric, so that some places relatively close in miles become distant whilst those distant in miles become close, depending upon which medium of communication is economically viable. The means of communication becomes as potentially determining of the incidence and quality of information flow as a fixed link is of the flow of goods and people. Lash and Urry (1994, p.25) list six media of communication from relatively cheap physical transportation, through wire cable (which can carry only small amounts of information), co-axial cable (capable of carrying images), microwave channel (good for person to person communication, as in mobile phones), earth satellite (excellent for reaching remote areas), to the expensive fibre-optic cable which excludes remote areas and privileges those sites connected to it. They suggest that the mode of communication has enormous impact not only upon 'time-space and time-cost convergence on a *global* scale', but also in 'processes of globalisation, localization and stratification' (p.25).

Obviously financial transactions are the major flow using fibre-optic cable and almost language-free communication, whilst poetry books travel by sea mail and depend upon literacy in the language of the poet. But between these two cultural extremes is the fascinating medium of satellite broadcasting. Countries which until recently could afford only limited local television stations, often restricted to government publicity and very old imported programmes and films, suddenly found themselves exposed to instantaneous communication from the major international broadcasting sources. Not only has world television news become available for those that can afford it, but also current programmes, recent films and international advertising; the world penetrates domestic space. It is obvious that the communication is only one-way, although some films may be dubbed into a local language, international languages predominate whilst images and symbolic systems are mainly American. Byron (1993) has demonstrated the popularity of uptake of satellite and cable TV in the Caribbean, but shows how the companies supplying it do so to the virtual exclusion of local broadcasting. The result has been a speeding up of the Americanization of Caribbean culture and a loss of local knowledge, which cannot compete with the sophistication of the international product (although we should not rule out the possible emergence of a counter-cultural movement to this).

National governments find it impossible to control much more than the licensing of particular channels, and day-to-day content becomes uncensorable. Clearly opinion about the political, social and cultural outcomes of this is a matter for debate. On the one hand it means that local political controls can be circumvented, so, for example, in Britain it is possible to listen to interviews with leaders of Sinn Fein on CNN from America, though these are not permitted on any British television or radio station, and, on the other hand, it is incontrovertible that political interpretations from American sources are becoming privileged throughout the world. This is not just a simple matter of propaganda, it is also a distortion of criteria of interest and relevance and, quite literally, point of view. When America orients itself towards the Pacific, rather than the Atlantic, it shows in the coverage as well as in the interpretation; Europe begins to see itself in a different light when it is mediated from afar. This is a new experience, but one which 'Third World' countries have long known, via BBC World Service radio broadcasting. It has long been recognized that many news events are manufactured performances for news broadcasts; the timing of an event takes account of media schedules at the communication core and world broadcasters are tipped off to be in place to record the 'something' which will happen. Wallis and Baron (1990) offer accounts of frightening cynicism on the globalization of American overseas military interventions, quoting the BBC's Kate Adie on the American bombing of Tripoli in 1986:

> Shortly before 2.00 a.m. in the morning [sic.] we got a call from the BBC in London saying that they had received a tip-off from the networks in Washington that something big was afoot – President Reagan had apparently asked for time on television. Then the raid started at exactly 2.00 a.m. Later it turned out that this was not what the Pentagon had wanted; they wanted a raid at 4.00 a.m. local time, when people are most drowsy, when the streets are empty, the time in the middle of a military shift. Why did it happen at 2.00 a.m.? Because that was prime time newstime in New York and Washington. . . . If you try to maximise the PR, at danger to your own men, that is a cynical use of it. (p.213)

When reports from Libya seemed less than eulogizing of the American position, there was immediate political pressure on the BBC, but Adie was unconvinced that this was prompted by a direct collusion between the Reagan and Thatcher governments. Instead she interpreted the action in terms of the force of networks of international media: 'people who own stakes in newspapers, who own parts of huge conglomerates, entrepreneurs who wish for a slice of the television audience in the UK . . . the BBC is the big animal they would like to slice up' (p.215). There are very real fears that as global broadcasting falls into fewer hands, news becomes more and more of a commercial product with large agencies covering the major international events and buying small and immediate items from small companies and freelancers. Whilst news then travels faster and more homogeneously, it also tends to cover only those locations which are within the view of the powerful news-brokers; there may be more news, but it tends to come from the same places. In 1986 CNN's 'Headline News' drew 77 per cent of its items from the USA, 6 per cent from Britain and Western Europe, 5 per cent from the Middle East, 3 per cent from Central and South America and 2 per cent from

Japan, leaving a mere 7 per cent for all the rest of the world (the then USSR got only 0.5 per cent of mentions!) This pattern is similar for all major channels.

But television is not just news and current affairs. Satellite TV means that local and national sport is eclipsed by that with an international spectatorship, reinforcing the creation of global superstars; it favours entertainment which can easily translate and as a consequence veers towards the bland and the spectacular. It is also impossible to miss the observation that Western notions of dramatic structure, morality and excitement prevail or, as Mattelart (1991) says, with only slight exaggeration, 'viewers in the South already know about dubbing, those in the North never will' (p.82). It must be a curious mix of elation at being in touch, and depression at exclusion, which informs one's attitude to television in those countries where virtually everything on offer is in translation and set abroad.

Mattelart is particularly concerned to demonstrate the way in which international media provides a cultural background to the marketing of international products. He observes that television programmes, whilst providing the context for advertising, also teach the symbolic structures which generate the desires that make global advertising meaningful. These contexts are Western*ized* as much as Western; enough local ambience is included to make the transformation dreamable, and a sprinkling of local presenters and accents help embed the global market in the local. The consequence is that famous, tiered, worldwide culture described by Saatchi and Saatchi in their 1985 Annual Report:

> Today's sophisticated marketers are recognising that there are probably more differences between midtown Manhattan and the Bronx, two sectors of the same city, than between midtown Manhattan and the 7th Arrondissement of Paris. This means that when a manufacturer contemplates expansion of his business, consumer similarities in demography and habits rather than geographic proximity will increasingly affect his decisions. . . . All this underlines the global approach. (quoted in Mattelart, 1991, p.53)

We can see that those who do not constitute a significant market are excluded from anything other than a peripheral role in this communication. They watch the performance and then they constitute a backdrop of local desire against which élites can project their achievable wants and enact a new enviable lifestyle. Not all viewers are owners, and non-owners become acutely aware of their information poverty and dependence; one of the desires is for a place on the Net. Access to information, whether we are talking of news, art or entertainment, becomes as important as access to consumer goods.

Yet it is not only lifestyles, goods and political attitudes which are purveyed by broadcasting, religions too are ethereally disseminated, along remarkably similar lines to soap operas. Stump (1991) demonstrates that a huge volume of full-time local and global Christian broadcasting emanates predominantly from the United States, transmitting the top five Protestant stations to virtually every part of the world; operating in more than 100 different languages in a massive propaganda drive. In addition to this, other stations also contain significant amounts of Christian programmes. He notes that such broadcasting makes relatively few outright converts, but succeeds in the task of revalidating Christian conversion in other contexts, such as missionary work. It also operates to create an awareness of Christianity into which personal contact can be slotted. Non-Christian religions rarely broadcast, the exception being Islam, in those countries where there is Christian media incursion and where programmes are used to strengthen resolve against possible defection from the faith; conversion is not the intention.

With international convergences and local divergences of the sort commented upon by Saatchi and Saatchi, with the virtual disappearance of some locations from the world view and from instantaneous communication, we claim that it is impossible to make any sensible generalizations about global (or local) cultures as if they were discrete bundles of traits. Ideas, commodities, ways of communicating are pressed into service in a multitude of different ways to construct flows of both meaning and nonsense which are not easily mapped onto the surface of the globe. Though we are clearly doubtful about the existence of a massive global culture, we are even less impressed by the possibility that there might be hermetically sealed local ones; there are only rhizomic flows of both global and local cultures.

SECTION II

A World of 'Others'

It is difficult to see the world except from one's own perspective and all of us are constantly falling into the trap of assuming that our own view is obvious, no matter how hard we try to think beyond our own egocentricity, ethnocentricity, or group solidarities. But as soon as we start to think about people who are not ourselves we lapse into the language of 'Othering' and, as one urges oneself to consider 'Others' or to see the 'Other' side of the question, those who are not like 'me' can start to slide into a homogenous mass of difference from 'me', *essentially* the same as each other. This is as just arrogant as the assumption that they are essentially the same as 'me'. It is implicit in the terminology that the self is taken as prior. The Other is secondary and the best that one can do for an Other is to extend a liberal tolerance, a condescension flowing from a benign superiority. Deleuze (1990) sees the construction of the Other as a means whereby one experiences oneself as an imagining being – arguing that the thought Other is a means whereby those places, things, ideas and sensations which are not directly experienced can be conceived of as having an existence, experienced by him/her. The act of constructing Others is unpleasant, but most of us, for lack of any better way to think, find ourselves performing it most of the time, trapped in the structures which help us to impose order upon the mass of variation that is the world.

Classification is an important part of logic. It occurs in both lay and scientific thought and can be seen as the way in which patterns of similarity and difference are generated through sorting, grouping and exclusion. But it should always be remembered that the system of classification is an artifice, there is nothing natural or obvious about it and it invariably serves to facilitate thinking from some perspectives more than from others. At the point of contact between categories there are things, places, people and ideas which are hard to classify because they fall between two or more of the available sets. Douglas (1966) has taught us to see that these things will be regarded as anomalous by the sorters, and from the anomaly there emerge ideas of sacredness, wonder, dread or revulsion. The things become *uncanny*, which literally means 'unknowable', but can also mean weird, frightening or unsettling. With differently constituted categories, the same things would not occupy the gaps but would become classifiable. Unconscious of their constructedness, few people confront their own categories and the gaps between them, though structures which are imposed from outside may seem appallingly artificial.

In this section of the book we work from the perspective of a number of different categories of people as constructed in Western society, but we are conscious that in doing this we are underwriting their constitution, even as we seek to deconstruct it. It is only if one makes the leap into poststructuralist styles of writing that it is possible to escape the rules which reinforce existing edifices and, even then, the escape is only partial, because of the constraints of language. The politics of communication are such that to make that leap would be suicidal for us in a book of this nature, though we shall give examples of *l'écriture feminine* which aims to step outside phallogocentrism by writing 'to get past the wall' (Cixous, 1991, p.5) whilst we remain timidly protesting within. This is always the bind of structuralism, it helps one think and communicate reasonably clearly, but it makes one think and say things one would rather not, unless one is fortunate enough to occupy the position from which the logic is constituted. It is for this reason that we need to familiarize ourselves a little with the enterprise of deconstruction, not because it is fashionable but because an understanding of empowerment and disempowerment depends upon an attempt to think outside the structure or 'against the grain'. One of the most important achievements of poststructuralist thought is to decentre, or disempower, the subject – the power which constructs meaning in the act of speaking. In examining and questioning the interestedness of the subject the constitution of her/his/its logical, moral and aesthetic categories is revealed. (But obviously there is no innocent position from which to conduct the interrogation, there are only political positions.)

In primary schools a great deal of emphasis is placed upon skills of verbal reasoning where one of the most important tasks is to select the 'odd one out' from a list of say 'lion, elephant, leopard, tiger' where the 'correct' answer is 'elephant' because all the others are cats, or is it 'tiger' because all the others can be found in both Africa and India, or is it 'lion' because it is the only one that is called the king of the jungle, therefore all the others must be seen as subordinates? Or is it a silly question because all of them are unique totemic beings with their own different characteristics in mythology? Most children learn that, whatever they might start off thinking about the uniqueness of things, it is even more 'wrong' to be unable to make some sort of classification than it is to make one which is deemed irrelevant. We are labouring the point because we insist that the structuring of language is actively taught, as is the logic of structuring. The child who seeks the odd one out on the basis of 'cuddliness' or 'frightfulness' is likely to find him or herself 'right' only occasionally and by coincidence.

Another standby of the verbal-reasoning exercise is the supplying of opposites. Children, when very young, are taught the principle of opposition (with no hint of the idea of mediation, which is itself part of oppositional thinking). Up/down; black/white; man/woman; sun/moon – whilst up cannot exist without down, the sun could easily be thought without the moon, and man and woman could be thought to be so similar that they were not worth opposing. What are we playing at when we generate logical systems like these? And how is one expected to think the difference between black and yellow? Yet opposition is privileged as the structural ideal of difference in much of modern society and an inability to structure one's logic in this way has long been penalized as lacking in rigour. Yet, as poststructuralist thought has been at pains to demonstrate, particularly in the work of the French feminists like Cixous (1991) and Irigaray (1987), the oppositions always contain within them an element of hierarchy – 'a' is conceived of as superior to 'not-a', which it defines in opposition. 'The word "woman" holds me captive. I would like to wear it out, to lose it, and to continue along on the trail of She who lives without this great worry' (Cixous, 1991, p.103).

It is for these reasons that we have grouped together the next five chapters which make up this Section; the categories exist with any meaning at all only because they are constructed by a subjective 'self'. We need to ask why gender, age, class and ethnicity are differentiating criteria and why we even have the concepts at all. It is all too easy to see the categories thus generated as inevitable or *essential*, rather than fabricated in cultural practice. They are classifications which exist in every social geography textbook but we want to question their status as things in themselves, whilst simultaneously recognizing that, having been constructed, they certainly have an existence in segmenting individuals from all around (see Box O).

These five chapters ruefully acknowledge that, from where we are standing, the world is logically structured for middle-aged, white, English-speaking, heterosexual, able-bodied, middle-class, scientifically educated, employed men. That is actually quite a tall order. It is,

BOX O Segmentation

We are segmented from all around and in every direction. ... We are segmented in a binary fashion, following the great major dualist oppositions: social classes, but also men–women, adults–children, and so on. We are segmented in a circular fashion, in ever larger circles ... my neighborhood's affairs, my city's, my country's, the world's. ... We are segmented in a linear fashion, along a straight line or a number of straight lines, of which each segment represents an episode ... forever proceduring or procedured, in the family, in the school, in the army, on the job. ... Sometimes the various segments belong to different individuals or groups, and sometimes the same individual or group passes from one segment to another. But these figures of segmentarity, the binary, the circular, and linear, are bound up with one another, even cross over into each other, changing according to the point of view. (Deleuze and Guattari, 1983, pp.208–9)

however, what is constituted as 'normal' in the society in which we live (though neither of us is it in entirety, neither of us has more than one attribute missing). We thought it would be a good idea to start with the problems that men have with this bundle of attributes and to consider some of the ways in which masculine frailties work themselves out in relation to spatialization of margins and centres, tame and wild land, and to think about the ways in which patriarchal structures constrain even those who are all too easily seen as benefitting from them.

All the other chapters focus upon ways of deviating from this norm, but we are trying to show how the differences which have been constituted as oppositional and hierarchical may also be seen as rich and equally valued. Here we encounter ideas about the weakening of boundaries which are talked of in terms of various hybridities, perversions, transgressions and subversions. The distinction made by Deleuze and Guattari between the smooth and the striated becomes important here – smoothness representing the continuity which can contain differentiable elements, striation representing the separation and structuring of difference. As Deleuze and Guattari emphasize, smoothness can at any time be deliberately striated, but so too can striations be smoothed, both occurring in response to desire and power: neither the categories nor an absence of categories can be regarded as anything other than temporary and not all differences make a difference (see Box P).

In this text we work from conditions of difference, isolating women,

BOX P Smooth and striated

Basing their distinction on the work of the French musician Pierre Boulez, Deleuze and Guattari (1988, p.477) argue that we live in basically two types of space-time in the world, State or striated space, and nomad or smooth space: 'In the simplest terms, Boulez says that in a smooth space-time one occupies without counting, whereas in a striated space-time one counts in order to occupy.' Striated space is overordered, segmentary and deeply progressive, like a map, it can give us our exact bearings and orientation. Smooth space holds continual non-segmentary and directionless variation. However, 'the simple opposition "smooth-striated" gives rise to far more difficult complications, alternations, and superpositions ... because the differences are not objective: it is possible to live striated on the deserts, steppes, or seas; it is possible to live smooth even in the cities, to be an urban nomad' (pp.481–2). The boundaries between the two are unclear, they shift continuously and even merge with one another in certain places as the smooth is striated and the striated is smoothed. What is crucial to each space are the ways in which it generates its own intensive qualities of liminality or borderlessness.

Williams (1993, pp.62–3) has applied the smooth and the striated to capital flows:

> The explosion in the production of striation has required machines dependent upon smooth spaces and the appearance of smooth space has necessarily accompanied striation. For instance, the need for ever increasing capital to finance states and companies has brought about vast flows of capital that move about the world at great speed; this capital occupies a smooth space and is very difficult to control and organise because it involves such great sums and because it can move so fast: a nomadic war machine has evolved out of the need for growth of machines of state.

those who are very poor, those who are conceived of as having a different ethnic identity, separating age groups, when we would sooner protest continuity from points of intensity. We are being conservative, even in this protest, for we have not given attention to the position which expunges the Other in its entirety as expressed most notably by Haraway (1991) in her refusal to separate human from machine or from environment in her eco-feminist view of the cyborg as the ultimate hybrid. Such a view evokes the personalized and kinship relations with that part of the world that 'we' bracket as 'nature' which are to be found in totemic systems of thought, where the sacred resides in the continuity. We have not, however, granted a chapter to nature, machines or, for that matter a great many other Others, such as those who are held to be sexually transgressive, or those defined as disabled. This does not mean, however, that we do not think that these constructions are any less interesting than those we have worked from and, to some extent, we have already thought about them in Chapter 4.

6

It's a man's world

Men can quite easily be shown up to be both ignorant and sexist. They have excluded women from the academy in a variety of ways: through the patronage of promotion, the rigid style of academic discourse and simply their claims to know. Much of humanistic geography was taken up with describing certain privileged *men's* feelings towards certain landscapes, or *topophilia* (Tuan, 1974). Rose (1993a) has shown that the idealizations involved in humanistic geography centred around an authoritative and aesthetic masculinity. A masculinity that once again through its universalizations silenced any articulation of sexual difference. Massey in her paper 'Flexible sexism' (1991) takes Harvey's *Condition of Postmodernity* (1989) and Soja's *Postmodern Geographies* (1989) to task for silencing women, through the use of unfounded generalizations and masculinist metaphors. She argues that this is revealed particularly in Harvey's analyses of the film *Bladerunner*, and his interpretations of Cindy Sherman's photographs. The thrill Harvey gets at playing the voyeur is all too obvious in his recent books, whilst Soja's penchant for monumentalism has not gone unnoticed (Deutsche, 1991; Morris, 1992). Tuan's, Harvey's and Soja's books are just three examples of the privileging of male vision, an authoritative, overpowering gaze. As Mulvey (1989) argued, this male gaze begins with scopophilia (pleasure in looking), a distanced, voyeuristic pleasure that projects its fantasies on to the female figure. We feel that perhaps it is time this male gaze was turned inwards (Middleton, 1992).

Instead of cataloguing all the instances where men silence women – a phenomenal task – we wish to critically examine this male gaze and its use in the construction of a variety of masculinities and femininities. All too often men talk of and for others without examining their own subjectivities. This is a point Rutherford makes in his book entitled *Men's Silences* (1992). The lack of a vocabulary for men to reflect adequately on their emotions may be one of the reasons why men continue to repeat the same phallocentric strategies. While there are a number of male vocabularies, particularly sexist and sexual ones in workplaces such as the

Stock Exchange (McDowell, 1993), male emotional vocabularies have had a tendency to be constructed negatively as feminine, camp and ephemeral, with repercussions for both femininities and masculinities. It may be time we opened up a new space in geography for a discussion of the construction of masculinities, and their role in the production of gender identities. In so doing:

> We must recognise the variety of gender identities that may be adopted by individuals of either sex, varying from place to place and time to time. We must be sympathetic to forms of masculinity and femininity that are not our own. We must be aware of the subjectivities of gay men and lesbians, as well as of heterosexuals of either sex, black or white, young or old. But we should avoid the trap of speaking only for other people and not also for ourselves. (Jackson, 1991, p.201)

But how do we avoid this trap of speaking *for*? Heterosexual men need to do more than just voice support for feminism and gay consciousness, for to do only that means to fall back into the same old gender stereotypes and representations. In this chapter I certainly don't want to claim that I represent other men, what I do want is for this chapter to be offered in the sense of dialogue. (In this chapter 'I' refers to Kevin's subject-position(s); 'we' sometimes means both authors, sometimes it is more inclusive.) However, this still has its problems as Middleton (1992, p.7) argues:

> Men, as well as women, have doubts about the very possibility of a radical discourse of masculinity. If men still have power denied to women how can these male oppressors produce an emancipatory political discourse on masculinity and subvert their own dominance? Isn't this more likely to be some kind of face-saving exercise than a radical political project.

Men's difficulties may be secondary to the problems women experience in articulating their powerlessness, but if we are interested in understanding how male dominance operates in societies, a critical investigation of different forms of masculinity must be undertaken so that men can be made aware of both their prejudices and differences. The oppression and violence constructed through dominant, hegemonic masculinities needs explaining if sexual spaces of fear and oppression are to be removed from cities (Valentine, 1992). Clearly, there are complex lines of power and desire relating patriarchy with different masculinities in different places. Masculinities are constructed and defined by their opposition to constructed femininities and in turn help to define femininities. Without each other they remain incomprehensible, but as Connell has argued: '"Hegemonic masculinity" is always constructed in relation to various subordinated masculinities as well as in relation to women. The interplay between different forms of masculinity is an important part of how a patriarchal social order works' (1987, p.183).

Cockburn's (1983) study of London print workers has shown how a hegemonic masculinity involved the subordination of young men as well as women, through the *rite de passage* of apprenticeship. But, whilst this subordination is tempered for heterosexual men, this does not seem to be the case for women or homosexual men (Connell, 1987: McDowell, 1989). But so that subordinated masculinities are not confined to a private realm and left unrecognized as an alternative, there is a need to take seriously the feelings, ambiguities, problems, differences and desires in that project of being a man, whether homosexual or heterosexual.

In what follows, I am not trying to speak for anyone but myself: yes, quite clearly this is an egocentric and personalized account, but if we want to speak clearly and critically about emotions and desires, it seems that we can but begin with our own. Within sociology, anthropology and cultural studies the critical re-evaluation of one's own emotions has been seen to be vital for understanding whatever processes we are studying (Okely and Callaway, 1992). In studies of masculinity, a critical autobiographical approach has recently been taken by a number of sociologists and cultural theorists, notably Morgan (1987), Jackson (1990), Lewis (1991) and Middleton (1992). The geographer Eyles (1985) has also experimented with this form of discourse.

In pursuing my autobiography, I do not wish to fall into the trap of telling various 'when I was a lad' stories. I want to interrogate my childhood imaginatively and critically, in order to expose 'the false unity of a stable, fixed, authorial identity' (Jackson, 1990). I want to examine a specific event in my childhood because as children most of us gradually conform to local socially acceptable patterns of *normal* masculinity and femininity. These dominant masculinities and femininities are continually changed and recreated throughout our life cycles, through each and every day, at home, at school and at work. For the sake of both reader and writer I intend to keep it short – I feel nervous about giving myself away totally. I will leave many questions unanswered but perhaps it may be a start for the critical articulation of a cultural geography of masculinities. After all, our reactions to different places are filtered through our gendered identities.

My friends say that the story I tell most is of getting expelled from a school I was attending in Germany when I was 16. In the past I have told it for masculine credibility with my male friends, and, stereotypically, for sympathy from my female friends; some friends I even think I told several different versions! I had just come back from a boarding school in Suffolk, where the discipline was almost intolerable. Time parameters were rigorously enforced, with lessons from 9.00 a.m. to 5.00 p.m., prep. (aration) or supervised 'homework' from 6.30 p.m. to 8.30 p.m. and bed at 9.00 p.m. Spatially, movement both inside and outside the school grounds needed written permission. However, the discipline began to crumble when the school was amalgamated with its sister school for girls. Nightly transgressions quickly became the norm as pupils of both sexes visited each others dormitories. Those caught were punished physically with a caning, mentally through detentions, or physically and emotionally through expulsion. But, desire ran its course and the transgressions continued.

Through its emphasis on a prefect system and sports such as rugby a masculine 'team spirit' was propagated. I felt the need to bond emotionally with my peers and spent a lot of time happily playing rugby until the day I broke my leg, rather unheroically in a practice game, and burst into tears. Jackson (1990, p.207) writes that sport, at first sight,

> seems a leisure pursuit, a casual letting off of steam at the margins of one's life. But looked at more critically sporting practices represent one of those apparently trivial but significant lifetime sites where masculinity is constructed and confirmed. Through regular, everyday actions, institutional practices, languages, images, daydreams and night-time fantasies and definitions, masculinity was actively produced in my body and in my head. Of course that process was not without its tensions and conflicts.

Far from being a neutral activity, the practice of different sports and games provides a spatial training of both the body and the mind in dominant forms of masculinity.

My emerging sexuality and academic identity was constructed largely in this period, but in some ways I reacted against the disciplinary laws and the excessively aggressive masculinity that the school attempted to instill through heroic history lessons, rugby playing and the cadet force (for example, I was the only boy to opt out of the Combined Cadet Force).

At this school a particular form of masculinity was on offer, one that emphasized physical courage, science, and a belief in hierarchical authority. Any relationships with women of all ages were strictly controlled (I remember being caught kissing a girl and having to face a wall for the best part of a Saturday), but in addition house-mistresses were peripheralized, female teachers were defeminized and 'phoning home to mum was frowned upon by the established male gaze as an affront to the school's masculinity: 'the achievement of manhood depended on a disparagement of the feminine without and within (Roper and Tosh, 1991, p.13). Numerous boundaries were set up dividing girls' and boys' activities, with the girls always being defined as the Other, their activities and desires were always marginalized. Despite this peripheralization of the feminine, the house-mistresses, matrons and girls were important figures for most boys, as 'stand-in' mothers.

After having done my O-level examinations at 16, I returned to Germany and went to another boarding school, expecting all the privileges of prefectorial status. However, this school didn't have a hierarchical prefect system, although it was just as firm in its discipline as my previous school in England. With no outlet for the masculinity I had constructed at the first school – no rugby and no prefect system – coupled with a sense of disorientation as an outsider in a new environment, I flouted most of the rules. I frequently went drinking outside the school grounds, mostly in parks. Eventually, the masculinity I had constructed at school in England, erupted into a petty act of violence, vandalism after a night of heavy drinking, and I was promptly expelled.

The production of masculinities involves the deployment of distinct notions of space and movement. We may see the inside of the school as rigidly structured and purified, the outside defined as structureless, dangerous and polluting. The spatial discipline of the school grounds and the exclusion of the feminine from within led to complicated feelings of insiderness and outsiderness. Inside the school a controlled, desexualized glorification of masculinity prevailed, but it was a masculinity bereft of power and autonomy. My transgressions, breaking the symbolic boundaries drawn up by the school, led to the punishment of permanent outsiderness – expulsion.

Clearly in the above narrative I have selectively edited out various other formative experiences in my life. I have also clearly spoken from my own ethnic and class bias, but as I said before I am not trying to speak for anyone else, my aim is only to show that constructions of masculinity are important in any understanding of subjectivity and thus for a critical, cultural geography. This doesn't condemn us to anecdotal material, it brings cultural geography alive. As Middleton (1992) argues, the paradox in using autobiographical material is that, the more self-critical the tone, the more virtuous the narrator may appear. However, 'this doesn't mean confessional writing is worthless. Far from it. Much of the impetus towards a better understanding of masculinity by men has come from just such confessional material' (Middleton, 1992, p.3). I would only advocate critical autobiography as a viable, qualitative approach in geography if it is used in conjunction with other methods, such as participant observation, in order to put the subject firmly back into geography.

One interpretative method gaining influence in geography at the moment is psychoanalysis. We would argue that a tentative geographical engagement with psychoanalysis, may perhaps offer individuals of different genders and sexualities a chance to begin formulating a radical politics of desire.

We relate to landscapes, and space in general, through the language that we use to articulate our emotions. Pile (1993, pp.136–7) has argued that 'psychoanalysis can describe and explain how power infects language and how language places people in relation to power . . .' It can also 'help articulate a politics of movement (and not merely position)', and 'respond to the demand for a politics of desire'. Clearly, what psychoanalysis cannot do is explain every facet of subjectivity. Moreover, psychoanalysis is no homogenous theoretical voice. Like geography it is composed of a number of competing schools, it is thus a starting space and not a completed project. Likewise femininity, masculinity, gender and sexuality, as concepts, are hard to grasp; they are also all unfinished projects, rather than simply ascribed roles. Much of what has been called poststructural theory reflects the uncertainties involved in coming to terms with these projects.

Gender may be thought of as the meaning of being either a man or a woman. It will always be locally constructed and there is no single masculinity or femininity. *Sex* refers to the biological differences between men and women, though there is nothing simple about what is attributed to nature and culture as debates about chromosomes and hormones in 'pathological' behaviour will testify. *Sexuality*, however, is another matter altogether, and has been the subject of furious debate ever since Freud, '"discovered" what every nursemaid already knew: the sexuality of children' (Mitchell, 1974, p.16). Historically, sexuality was explained by the nineteenth-century sexologists such as Havelock Ellis through biology. Freud contributed to these ideas by suggesting that masculine and feminine identities, although not determined by biology, are probably the result of the ways in which boys and girls interpret their own physiology and relate sexual instincts back to biological sex (Valentine, 1993). More recently sexuality has been interpreted by psychoanalysts such as Lacan (1977, 1982) as a social *construction* at the linguistic or symbolic level, in a critique of the Freudian's *essentialist* claims (see Box Q). The geographer Valentine (1993, pp.238–9) has argued that sexuality and gender identities cannot be separated: 'socially constructed and interlinked gender and sexual identities are not fixed but are constructed and reconstructed over time and space.' Yet, 'the temptation is still to see sexuality as interpersonal sexual relationships, and sexual phantasies [*sic*] or auto-eroticism as perverse. ... In other words, sexuality, to be "decent", must have an object' (Mitchell, 1974, p.16). That object of desire has been termed the 'phallus', which again should not be reduced to the biological differences between the sexes but needs to be thought of as a *linguistic signifier* for the fundamental division between the sexes (Lacan, 1982).

Lacan (1977) argued that the Self is constructed in relation to Otherness. Important in this process of identification is the 'mirror stage', the moment when a child sees itself in a mirror, or in the reactions to its actions by another, and *mis*recognizes that it is a bounded body and gains pleasure from this. From that moment the child's gaze 'is always torn between two conflicting impulses: on the one hand, a narcissistic identification with what it sees and through which it constitutes its identity; and on the other, a voyeuristic distance from what is seen as Other to it' (Rose, 1993, p.103). In phallocentric societies this process of identification involves a privileging of the active male gaze over images of nature and landscape in the construction of the metaphor Mother Nature.

As children we begin to fragment our notion of a bounded self as we move from the realm or space of the real, through the imaginary towards the symbolic: 'Each of these is always present, but individuals are located differently in relation to each of these spaces, we are placed but we are

BOX Q Essentialism

Constructionists tend to argue that there are no essential biological or innate characteristics in order to stress the differences, real, imagined or symbolic, that exist between individuals. Essentialism is typically defined in opposition in order to lay claim to certain commonalities between individuals. But, as Fuss (1989, p.xii) has argued, 'the binary articulation of essentialism and difference can also be restrictive, even obfuscating, in that it allows us to ignore or to deny the differences within essentialism'. In the hands of a hegemonic masculinity, essentialism can be a very powerful tool; Harvey (1989), Soja (1989) and Tuan (1993) all make use of essentialist arguments, whilst the extreme case of Appleton's (1975, 1990) biological determinism of the symbolism of landscape should make us extremely wary of making essentialist gestures. But, whilst we have tended to avoid positing generalized, essential categories in this book, there is a case for arguing that 'a provisional return to essentialism can successfully operate, in particular contexts, as an interventionary strategy' (Fuss, 1989, p.32) for non-hegemonic groups, a tactic that Spivak (1990) has chosen, for example.

irrevocably lost' (Pile, 1993, p.135). Central to our movement through these realms, and to the production of our ambivalent masculinities and femininities is the gendered ordering of our formative relationships with our parents. Lacan argues that the experience of the father is important, as an infant identifies him as possessing the *phallus* – 'not a symbol of patriarchal authority, but a symbol of plenitude and completeness which the infant perceives the mother is lacking' (Rutherford, 1992, p.146). For Lacan the phallus is the fusion of desire and language – 'an elusive quality; a desire to possess one's own desire' (1977, p.148). Hence,

> Masculinity and femininity are differences articulated by something imagined to be absent from each. ... The meaning of female sexuality is determined through its lack in relation to the phallus. But it is the female body which is its original source. The significance of this contradiction is that once the small boy discovers that his father also lacks the phallus, he is confronted by a lack which cannot be filled. For the father is also unable to provide a meaning of this lack. ... The father doubly lacks. (Rutherford, 1992, pp.148–9)

Parents thus play an important role in the structuring of men's feelings and their sexual desires.

A school of film theory, based around the ideas of Lacan has developed largely within the pages of the journal *Screen*. Within geography, the interpretation of films in order to show gender–environment relations has recently risen to prominence with the work of Aitken and Zonn (1993). The analysis of both the internal dynamics of films and film genres as well as the actual consumption and identification of characters, landscapes and relationships in films has led to substantial critique of the masculine, voyeuristic, oppressive gaze in cinema. In her seminal paper 'Visual pleasure and narrative cinema', Mulvey (1975, p.18) pointed out that:

> Going far beyond highlighting a woman's to-be-looked-at-ness, cinema builds the way she is to be looked at into the spectacle itself. Playing on the tension between film as controlling the dimension of time (editing, narrative) and film as controlling the dimension of space (changes in distance, editing), cinematic codes create a gaze, a world and an object, thereby creating an illusion cut to the measured desire.

Again, I would like to turn this gaze inwards and briefly explore three films which examine father–son relationships, and the construction of masculine desire.

Peter Weir's (1989) film *Dead Poets' Society*, examines how masculine characteristics are constructed for boys through the organization of relationships at school, around the phallus and its lack. The 'dead poets' stand for a naturalistic creativity that a group of boys discover through the arrival of a new English teacher at their highly disciplined and conformist school. Poetry, especially Whitman's, becomes a metaphor for their emerging sexualities and masculine identities, and particularly their relationships with their fathers and other male authority figures such as the new English teacher Mr Keating. The use of narcissistic close-ups throughout the film serves to emphasize the closed space of the school, as well as the boys' emerging identification with both their real fathers and their symbolic 'good' father, Mr Keating.

The film begins with an argument between a boy named Neil and his father about his fate versus his desires – his father demands that he drops his involvement in the school magazine so that he can concentrate on his studies. His friends' lives have also been similarly predetermined by their fathers. As Rutherford (1992, p.158) advises us (following Lacan, 1977): 'their lives have been mapped out by their fathers. Hope, pleasure and desire are subordinate to their incorporation into the Name-of-the-Father.'

The boys discovery of poetry is coupled with their emerging interest in their own sexual desires, but it is also associated with a distinct space, namely the marginal location of the cave in the wilderness, where they meet. The group of boys take Keating's book of dead poets' poems down to a cave one night and hold the first meeting of the Dead Poets' Society. One boy recites a poem, another tells a horror story, and another a rhyme in which a woman is killed – a symbolic mother. This rhyme is written on the back of a pornographic female image, to symbolize the boys' heterosexual desires. However, the poetry used in the film is largely homoerotic, for example, Whitman's (1912) *O Captain! My Captain!*:

O Captain! my Captain! our fearful trip is done,
The ship has weathered every rack, the prize we sought is won.

But O heart! heart! heart!
O the bleeding drops of red,
Where on the deck my Captain lies.
Fallen cold and dead.

O Captain! my Captain! rise up and hear the bells:
Rise up – for you the flag is flung – for you the bugle trills.

Here Captain! dear father!
This arm beneath your head!
It is some dream that on the deck,
You've fallen cold and dead.

My Captain does not answer, his lips are pale and still.
My father does not feel my arm, he has no pulse nor will.
The ship is anchor'd safe and sound, its voyage closed and done.
From fearful trip the victor ship comes in with object won:
Exult, O shores, and ring, O bells!
But I with mournful tread,
Walk the deck my Captain lies,
Fallen cold and dead.

Despite there being a question mark over the corpus of Whitman's work and his democratic ideals, ethnocentrism and style (Simpson, 1990), in this poem the images of a son looking up romantically to his father are subverted by the son's quest for his own destiny, his own identity, for which he must metaphorically 'kill' the figure of his father and thus 'win' the object of desire – the phallus – which is also the castrating apparatus that wounds both father and son (Deleuze and Guattari, 1983). The spirit and cycle of this poem continues throughout the film (Rutherford, 1992).

The boys continue to transgress, but not break outright, the rules of the school under the auspices of Mr Keating. Whilst Keating gives a lesson on the dangers of conformity they walk around the school courtyard, unknowingly under the hierarchical gaze of the headmaster. Keating also attempts to subvert the traditional engendering of masculinity through sport. He gets the pupils to read a poem before they kick the ball, to the strains of Beethoven's *Ode to Joy* – ironically based loosely on a romantic poem by Schiller that sought to emphasize the 'manly' values of courage, truth and pride.

The theme of desire versus the Law, and the Name of the Father continues as Neil transgresses his father's demand that he should not take part in any further extracurricular activities, by acting in a production of Shakespeare's play *A Midsummer-Night's Dream*: 'In his role as Puck, Neil personifies everything that his father is not, fluidity versus rigidity, movement versus stasis, desire versus the Law, feeling versus its lack and the inkling of a destiny overcoming fate' (Rutherford, 1992, p.168). Neil's father finds out about his son's involvement in the play, and comes to watch. Under his father's gaze of authority, Neil, as Puck, the nature sprite, seems effeminate, and he represents Otherness. Neil's father takes him out of the school and intends to send him to a tougher military academy, in order to punish him. His mother looks on, entrapped by her feelings towards her son and her relationship with her husband. That night Neil, feeling trapped and worthless, and in an insecure *place* (the double lack) takes his father's gun and commits suicide. We can recall Whitman's poem in which a man lies dead, whilst his best friend laments. But the 'cycle' is not complete as a witchhunt is conducted to punish Keating for his role as the 'good father' and the Law comes to bear. The film shows the predicaments of young boy's emerging sexualities through its use of both heterosexual and homosexual imagery – a practice common in mens' magazines such as *GQ* and *Arena* (Rutherford, 1988). The boys' sexualities are connected emphatically with the geography of the school (see Chapter 11). The enclosed space of the school or home represents phallocentric power, whilst open space and the cave in the wilderness, represent escape and transgression.

Other films, such as Stone's *Wall Street*, also take this theme of a father–son relationship in the engendering of stereotypical masculine characteristics of greed, aggression and power in a ritual process. McDowell (1993) has recently explored the creation of sexist vocabularies and subjectivities of business*men* working in the City of London. The use of phallocentric metaphors such as the 'Big Bang' and the imagining of clients as women 'to be screwed', is an example of how the dominant male gaze works through language. We can begin to understand the construction of this aggressive masculinity through the insights of psychoanalysis once again. The film *Wall Street* portrays the hard-hitting almost exclusively male world of the New York Stock Exchange, and the film could be simply interpreted as a moral tale revealing the level of corruption – insider trading – in the business world (Denzin, 1991). But it also shows the negotiation of masculine identities in a specific workplace, and the transfer of male power through the lineage, a process fundamental to the reproduction of patriarchy.

The theme of the movement of the phallus from father to son is also found in films given historical settings, such as *Sommersby* (Amiel, 1993). This film is based upon a French folk tale and film *The Return of Michel Guerre*, and we can perhaps begin to see the phallocentric theme as core archetype of Western societies. *Sommersby* is set against the background of the end of the Civil War in the United States. We again have a father–son relationship, and as in most mainstream films, women are given only supporting roles – catalysts for male heroic endeavours (Aitken and Zonn, 1993).

A man claiming to be Jack Sommersby returns from the war to 'his' home town in Tennessee and after all the rejoicing he is reunited with 'his' wife, Laurel, and 'their' son. Despite a physical resemblance to Jack Sommersby, the man is an imposter called Horace Townsend, and has a very different character from Sommersby, who was violent and racist. The new Jack is a symbolic 'good father' and husband; possessing the phallus, and completing the holy family. The 'wife' realizes he is an imposter but connives and enters into a sexually transgressive relationship with him. He establishes a successful tobacco plantation by leasing his land to local men and women, black and white. The community come not only to admire and believe in him, but also to have an economic stake in him. The real Jack Sommersby is wanted for murder and the sheriff comes riding in, embodying the Law, to break up the family and community.

At his trial he clings to his role as the 'good father' and legitimate husband, despite his 'wife's' attempts to prove him to be an imposter. Knowing that the community would lose their land if he admits that he is not their legal landlord, and not able to lease the land, he desires against his own interests for the benefit of the community; for a place that has become his home, for the sake of his 'son' (who testifies that he is the *real* Jack Sommersby), and for the sake of his masculinity. He is found guilty and hanged. A man lies dead once more, having transgressed the Law, in search of the object of his desire.

As mainstream films, *Dead Poets' Society, Wall Street* and *Sommersby* reproduce dominant Western forms of masculinity, but our identification with the characters of films is not a simple process:

> identification is never simply a matter of men identifying with male figures on the screen and women identifying with female figures. Cinema draws on and involves many desires, many forms of desire. And desire itself is mobile, fluid, constantly transgressing identities, positions and roles. Identifications are multiple, fluid, at points even contradictory. (Neale, 1983, p.4)

Mulvey (1989) argues that enjoyment of films re-enacts the 'mirror' stage of child development and forces us to identify with the *hero* perspectively organized in relation to the landscape. As we grow up it seems as if we are bombarded by a steady stream of images and representations of masculinities and femininities, which we interpret and negotiate as part of our gender identities (Chapman and Rutherford, 1988; Craig, 1992).

We can see the same themes of dominant forms of masculinity repeated in a whole range of novels and films, as well as in expressions of popular culture such as travel, fashion and rock music (Mort, 1988; Rutherford, 1988; Nicol, 1990). In Lowry's autobiographical novel *Ultramarine* (1933) we find a concern for manhood translated into a *rite de passage* through a boundless sea voyage on board an aptly named ship *Oedipus Tyrrannus* (Porteous, 1987). Cresswell (1993) has recently examined Kerouac's novel *On the Road* (1957), and outlined how mobility is used by subcultures (in this case the Beat Generation) as a form of ambiguous and contested resistance. However, Cresswell notes that transgressive movements like the Beats often reproduce many of the established norms of the day, in this case the dualism of public (male) and private (female) spaces: 'In *On the Road*, travel in space is connected with masculinity while place and home are feminine' (Cresswell, 1993, p.258) – a stereotypical dualism of modernity. In another of Kerouac's (1959) autobiographical novels, *Maggie Cassidy*, we can see the familiar construction of masculinity in a group of boys through baseball. I feel that it is in the poems of another Beat writer, Allen Ginsberg that we get closest to the contradictions of sexual desire and its lack, the search for the phallus and the engendering of dominant masculinities, but also the idea of a rhizomic line of flight:

> the Green Automobile
> which I have invented
> imagined and visioned
> on the roads of the world
>
> ———
>
> . . . we'll be real heroes now
> in a war between our cocks and time:
> let's be the angels of the world's desire
> and take the world to bed with us before we die.
>
> Sleeping alone or with companion,
> girl or fairy sheep or dream.
> I'll fail of lacklove, you satiety:
> all men fail, our fathers fell before.
>
> (Ginsberg, 1963, pp.11–15)

Resistance to dominant masculinities can take many geographical forms, however, not just through mobility. A possible regaining of the construct of 'home' as a feminist site of resistance has been examined by bell hooks (1990). The imaging of Amsterdam as a gay capital in travel guides serves as another form of resistance (Duyves, 1992), as does the sexually transgressive experience of *Fucking Among the Ruins* (Binnie, 1992). These forms of resistance tend to involve a problematizing of the public- private-space dualism (Bell, 1993).

Black men's resistance to dominant masculinities has taken a variety of forms (Mercer and Julien, 1988, Westwood, 1990; Gilroy, 1993). As Kobena Mercer and Isaac Julien (1988, pp.138–9) argue, 'black people have entered sports not just for their own individual gain; by using their public status, they have articulated a political stance', whilst, 'It is in the medium of music – always associated with dance and the erotic potentialities of the dance floor – that black men and women have articulated sexual politics'. But this is not to say that this resistance is in any way unproblematic and that rhythmic dance is naturally a black attribute. Historically, in Britain and the United States the relationship between dominant masculinities and ethnicity has involved complex processes of imagining and projection of a whole series of supposedly natural attributes, such as laziness. These discourses were then used in racist, colonial theories of expansion (Walvin, 1987; Boyd, 1991; Dawson, 1991; Jackson, 1991; Jenkins, 1992; Mangan, 1992). And, as Jackson (1991, p.205) argues: 'black men are constantly represented as objects for the white viewer's gaze, with a powerful mixture of motives from sexual exploitation to homo-erotic desire.'

If we are to understand the geographical implications of a whole host of cultural and political phenomena, geography will have to take the study of masculinities much more seriously than it has done so far. Central to this task is a deconstruction of the normal male gaze, a heroic gaze that voyeuristically constructs monuments in landscapes, an issue we consider in more depth later in Chapter 13. Masculinities are in a perpetual condition of achievement, assertion and negotiation, but they are never securely possessed. With criticism they can change, and we may then see a shift in the dominant power structures in Western society.

7

Room for women

There is a widespread tendency to associate women with the domestic sphere in opposition to the public realm of men, but there is a very real sense in which many women do not feel at home anywhere. Irigaray most eloquently catches a common relationship between woman and 'home' in her paper 'Sexual difference' (1987), when she writes of man's nostalgia for his foetal dwelling place resulting in:

> the endless construction of substitutes for his pre-natal home. From the depths of earth to the vast expanse of heaven, time and time again he robs femininity of the tissue or texture of her spatiality. In exchange, though it never is one, he buys her a house, shuts her up in it, and places limits on her that are the counterpart of the place without limits where he unwittingly leaves her. He envelops her within these walls while he envelops himself and his things in her flesh. (p.123)

Irigaray's incarcerated but invaded women can be found in many fictitious and theoretical works and certainly not all of these are written by active feminists. The fact that so many women find themselves both installed in and acting as someone else's home can make them homeless:

> The mother woman remains *the place separated from its 'own' place*, a place deprived of a place of its own. She is or ceaselessly becomes the place of the other who cannot separate himself from it. Without her knowledge or volition, then, she threatens by what she lacks: a 'proper' place. She would have to envelop herself, and to do so at least twice: both as a woman and as a mother. This would entail a complete change in our conception of time and space. (p.122)

The association between women and men's homes has often led to an identification in feminist geography between women and place, rather than space, an assumption that women operate in the realm of the particular and the local and have difficulty extending their view of the world beyond the concretely experienced. So, for example, we find Bondi (1992) interested in the way in which the aspirations of women transform the fabric of urban locations through their own preoccupations with lifestyle in a process of gentrification, and Bondi and Peake (1988) comment that it is only at the level of community (i.e., local place) politics that women and women's considerations are prominent. Rose (1993a, 1993b) eloquently demonstrates the way in which humanist geographers feminized place by associating it with a personalized essence which slid easily into idealized woman. Women as located, men as identifying; women as passing from place to place, men as moving through space. She comments that 'women are never theorised as relaxed enough, satisfied enough or powerful enough to feel "at home in place"' (1993b, p.56). Massey (1993) also rails against this fixity and feminization of place and demands that we think through the easy way in which place is often characterized as feminine in contrast to space which is masculine, claiming that similar processes of space-time compression impinge upon women's and men's daily lives and that to see women as in some way especially grounded in the local as an extension of the home is a sophistry.

The association between women and the domestic sphere is to some extent based upon a tautology, because we find geographers and ethnographers throughout the world classifying the domestic realm as the places where women and children are permitted and the public realm as the places where they are wholly or partially excluded. (So, for example, schools for very young children are called *écoles maternelles* in France and nursery schools in Britain, sliding them conceptually out of the public into the domestic realm.) In much of Africa we find the domestic realm extending to the cultivated fields where women work, but not necessarily to the space in front of people's houses where men talk. In those Muslim households where a strict purdah is maintained, parts of the house where visitors are received may be seen as semi-public and the domestic realm shrinks down to a *zenana*. The very opposition of the constructs domestic and public is itself more illustrative of phallogocentrism than of the actual utilization of space. We need to be conscious, however, that these terms 'domestic' and 'public' are frequently imposed by ethnographers, rather than unproblematic translations of indigenous terms since there is often no local concept of the difference *other* than in gender terms, hence the caution about tautology. Bearing this in mind, we can still say that where there is a conception of the divide between public and domestic, women are nearly always a vital ingredient of ideal domesticity.

Adolf Loos, a prominent modernist architect, famous for his domestic buildings, was in no doubt that he designed for 'gentlemen'. Colomina (1992, p.74) reports Loos as explaining to Le Corbusier his fondness for windows which were translucent rather than transparent, saying that 'the cultivated man does not look out of the window; his window is a ground glass; it is there to let the light in, not to let the gaze through'. He justifies his position by stating that the cultivated man can come and go as he pleases between his home and society. For Loos, though, the householder's womenfolk explicitly act as guardians of the inner sanctum that is *his* home and he is famous for designing a house with an internal drawing room which surveyed other rooms, like the dining room and music room, but which did not give onto the outside world at all. This room he called the '*Zimmer der Dame*'! His *wife's* bedroom in his own flat (which he designed and referred to as a 'bag of fur and cloth') was dominated by a huge double bed, the base of which rose fur-covered and continuous out of the same fur which covered the floor. The walls were hung and muffled with heavy cloth, and the external view was excluded. No room could have been more explicitly the womb, the 'foetal home' that Irigaray was describing. Clearly it was impossible for Lina Loos to *be* herself in that room. Her fantasies could never have been expressed in what was ostensibly her own private space; she *existed* purely as an extra in Loos' dwelling.

It might be easy to dismiss Loos as a fetishist, but his is only an extreme form of what many people still consider an acceptable representation of the relationships underpinning the domestic, an emphasis upon the interior and the exclusionary and an explicit retreat for the male head of household. Although it is women who are charged with the role of homemaker, what they are required to make is often the product of masculine desires, within the context of an architecture which, until recently, has taken little account of women's preferences. Girouard (1978) has shown how, with the rise of middle-class gentry, the Victorian English country house began increasingly to separate male and female activities. A separate door from the garden into an exclusively male part of the house equipped with smoking room, billiard room, library and gun room allowed the men to come and go without being easily intercepted by the women in the drawing room and morning room. The women's only personal space was in their bedrooms. Gradidge (1980) has also demonstrated that in scaled-down form, these aspirations for male seclusion within the house continued through the early years of the twentieth century in studies incorporated into relatively modest middle-class houses.

In these grand houses women were on display, installed in settings which mirrored the status of their owners. Their coming and going was monitored, as were their callers. It is little wonder that so much early feminist writing, produced by leisured women in just such circumstances, should focus upon the fantasy of escape from the 'ideal' home into a less opulent 'home of one's own'. Chopin's *The Awakening* (1984) is a brilliant example of the genre and one which is particularly apposite for cultural geographers, so rich is it in metaphors of place, space and environment. The book was regarded as completely immoral when it was first published in 1899 and it still has a power to shock in that Edna, the central character, so relentlessly pursues her own romantic and sexual desires, casting off not only her family home in favour of a tiny house of her own, but also rejecting her husband, children and ultimately her own life in her bid to be free. Edna had discovered that in the role of a respectable wife, she and 'her' house were perpetually at the disposal of her husband's business interests; that, in the town where her husband had a reputation to guard, there was no place inside or outside where she could be at ease. At first she withdrew into despair and listlessness, then escaped into recklessness, finally swimming out into the sea at the holiday resort where she had known freedom.

There is an apparent confusion in the symbolism of the room in women's conceptualizing of space. Rooms can serve as bolt holes or prisons; women try to escape into and out of them. They can represent whole realms of personal autonomy or enforced gregariousness or exclusion. Sand (1929, p.24), lamenting the end of an affair, wrote in her diary: 'I go out. I seek distraction. I try to rouse myself, keep going. But when I come back to my room in the evening I go crazy . . . I do not want to stay at home in the evening. I would rather play dominoes in a café than spend an hour after dinner in my own house'. Duras (1990) commented that, whilst practically all of the women she knows have read Woolf's *A Room of One's Own* (1945), almost none of her male acquaintances had done so. It is one of those special books which seems so *obvious* when it has been read, but which changes one's perspective if one is a woman. Woolf shows that without a small income of her own and a room from which she can exclude the rest of the world, a woman can rarely achieve artistic creation. This is not just a practical matter of exclusion of the noise and demands of others but a question of groundedness and self-assurity. The room of one's own is not under the control of the head of household – it does not display his taste and social standing to visitors. From the room of her own Woolf sallies forth alone, like Loos' cultivated man, into the public realm. Very few women have a room of their own (the *Zimmer der Dame* certainly wasn't one, neither was Lina Loos' bedroom). Those of us fortunate enough to have one will attest to the fact that, for all that it is usually more spartan or shabby than the rest of the house, this is the room that female friends express the greatest desire for. One is seen to be so 'lucky' to have such a room and the obvious (and true) interpretation is that one has an unusual status in one's household. Some women appropriate the kitchen as a space of female sociability and creativity (not just cooking, but writing or painting), but the tendency towards larger living room/kitchens in middle-class households and tiny fitted food-preparation workshops-cum-laundrys for everyone else takes even this semi-private space away. Whilst the whole house represents the private realm for a man, even if he does then crave an even deeper inner sanctum of a study or workshop, for most women with a family this is not the case. For her the home is the locus of housework, which either constitutes a major part of her identity or serves as a 'second shift' after paid work. Though a considerable pride is often taken in the house, there is little opportunity to escape from the uncompleted work it requires.

Yet so often the woman living alone is represented as defective. Betjeman's (1954) poem about 'Business Girls' living alone in rented rooms in London describes them as leading uneventful and beleaguered lives:

> Rest you there, poor unbeloved ones.
> Lap your loneliness in heat.
> All too soon the tiny breakfast.
> Trolley-bus and windy street.

Lispector (1986, p.14) works though the idea of a virtually meaningless female life by constructing a sparse tale about a typist in Brazil:

> There are thousands of girls like this girl from the North-east to be found in the slums of Rio de Janeiro, living in bedsitters or toiling behind counters for all they are worth. They aren't even aware that they are superfluous and that no one cares a damn about their existence. Few of them complain and as far as I know they never protest, for there is no one to listen.

The single women thus conceived of as residual to ordered society cannot offer any hope of freedom and her place of residence is seen as marginal to ideas about home.

But a room of one's own achieved by one's own efforts is a minor rebellion against conventional domesticity. It hardly needs saying that this rebellion is accessible only to those women who live in relatively affluent households which are not overcrowded, but even when there is abundance (as in the country houses referred to earlier) it is not often seen as sufficient to establish women's space: there are plenty of houses which have numerous 'public' rooms, rooms for the children, a study or a workshop, but no room for the woman.

Whether or not a woman's 'place' is in the home, a great deal of female effort goes into home-based activities and the rewards for this are not easily measured, no matter how real they may be. Although women now expect to undertake paid work, whether full or part-time, home-based or outside, we rarely find the economics of the labour market imported into the domestic sphere, where an almost completely different value system operates, based around such things as love, duty, self-esteem, guilt and fear. In this economy very long-term credit is extended and accounts are not kept in money, but they *are* kept and the notion of indebtedness is invoked when things do not run smoothly. Resources get very confused and one reason why the home-economics manuals of the 1950s seem so hilarious today is that they seek to apply Fordist ideas about production to a situation which is inherently unruly. At that time the language of advertising and women's magazines became 'scientific' and 'utilitarian' – new products like detergents were endorsed by 'scientists' (Friedan, 1966). Suddenly 'labour'-saving devices were on offer, despite the fact that no one had ever talked about 'labour' outside the market economy before (other than in childbirth).

This new way of rationalizing and modernizing the home and its activities went hand-in-hand with a new suburbanization and an era of new town development. This meant a considerable rise in the standard of housing for a great mass of the population but a simultaneous clearing of inner-city housing and a consequent disruption of local communities and urban extended kinship. McDowell (1983) elegantly demonstrates the connection between suburbanization and assumptions about the importance of the nuclear family. She argues that suburban development could not initially have taken place without the emergence of a concept of the independent nuclear family with a division of labour in which the roles of husband and wife were clearly demarcated – the man working for wages outside the home, the woman acting as homemaker. Suburban settings and new towns built on 'garden city' principles were seen as ideal places in which to bring up healthy families and as sites of personal improvement. Men commuted to work in urban centres. Women, isolated in suburbs and urged on by processes of modernization, aspired to higher standards of domestic provision and efficiency and, to finance these, began to look for part-time work, close to home. The desire to own one's own home was fostered by the idyll of suburban living and the advertising of speculative builders playing upon notions of family solidarity and social mobility. To finance this, women were increasingly drawn into the labour market. Partly in response to the presence of a relatively cheap, compliant part-time female workforce, light industry and retailing were then attracted to suburban districts, enabling women to work whilst maintaining a significant presence in the home.

This relatively straightforward account of a shift in domestic life at the height of industrial modernism has much to commend it, but it is only recently that the academic world has begun to focus upon the detail of women's lives in the suburbs and what is now emerging reveals complicated attitudes to home, paid and unpaid work. Attfield's (1989) work on domestic interiors in the new town of Harlow in the 1950s shows that women were only partially absorbed into the modernist project. Although designers expected them to live new modern lifestyles in houses which were expressly designed for that purpose, the women found themselves ambivalent about this. Attfield demonstrates that women were proud of their new modern houses. They, rather than their husbands, were happy with non-traditional layouts and utilitarian workspaces and remembered how excited they had been when they were able to afford their first refrigerators and washing machines. But they also furnished the 'public' parts of their homes in non-modernist styles, softened picture windows with heavy curtaining and pelmets, replaced functionalist heaters with fake fireplaces and generally did all sorts of things which made arbiters of 'taste' throw up their hands in horror. They cluttered up the clean lines of modernism which they could accept in their own workplace of the kitchen but not in the recreational space of the 'lounge' (which they worked at producing and maintaining). If we think back to Loos' version of functional modernism, where he saw the *function* of the house as being to turn inwards in order to protect the man in his retreat from the world, we see the specialist suburban housewife making the same provision. Partington (1989) has looked at the design decisions of 1950s' women, and has found the same ambivalence to modernist design – particularly desirable when the designer has erroneously assumed a single function for something which is perceived by the woman to have multiple uses. For example, there was considerable dislike for merely 'functional' heaters which could not offer the same focal point and display space as a fireplace.

We find the same ambivalence about the newness of the town. There is a pride in its Utopian modernity, in it being a first 'proper' home, but a horror of its desert-like qualities and of personal isolation. It is important to see that as families moved to new towns and suburbs, they did not just transplant old ways of living and use these to subvert the intentions of planners, as they were often accused of doing. Neither did they buy into the approved form of modernist living. Instead we can see women responding to new spatial experiences in a creative way, learning to make their own adjustments to modern space in ways which were only selectively modernist. They were acting out the old split relationship to home which oscillates between a source of identity and a site of angst. If the angst has been stressed in the representations of 'suburban blues', it is because new locations offered so few opportunities for women outside the home. The domestic became everything, and for many women, it was found less than wholly desirable.

The same processes of modernization and suburbanization were experienced in all of those parts of the world coming under direct imperial control. Wilson (1991) shows how in the newly designed Lusaka (Zambia) of the 1950s, garden-city principles prevailed, such that, although there was a central business district and an administrative district, all residence was in suburbs. African women were almost completely excluded from the European residential districts and from the centre. They had few employment prospects and there was little productive work they could do in the African-designated suburbs. They were encouraged, administratively and by economic pressures, not to reside in town at all. After independence, Lusaka began to accumulate unofficial housing in shantytowns, but infilling and the suburban principles continued for more élite groups. Women began to participate more fully in formal and informal economic activity, coming into the towns in increasing numbers. They certainly did not regard themselves as dependents of men, but found it very difficult to obtain accommodation without at least temporary male support (Schuster, 1979).

It is easy to assume that the feeling of being confined by the domestic realm is a phenomenon of modern life and that in traditionally organized societies women are more closely attuned to their roles and statuses. However, if we look at the girls' rituals and symbols of puberty in many traditional societies we find an attempt to come to terms with this schizoid relationship with a marital home, and by implication, the rest of the world to which it stands in contrast. Figure 7.1 shows a plan of a rural

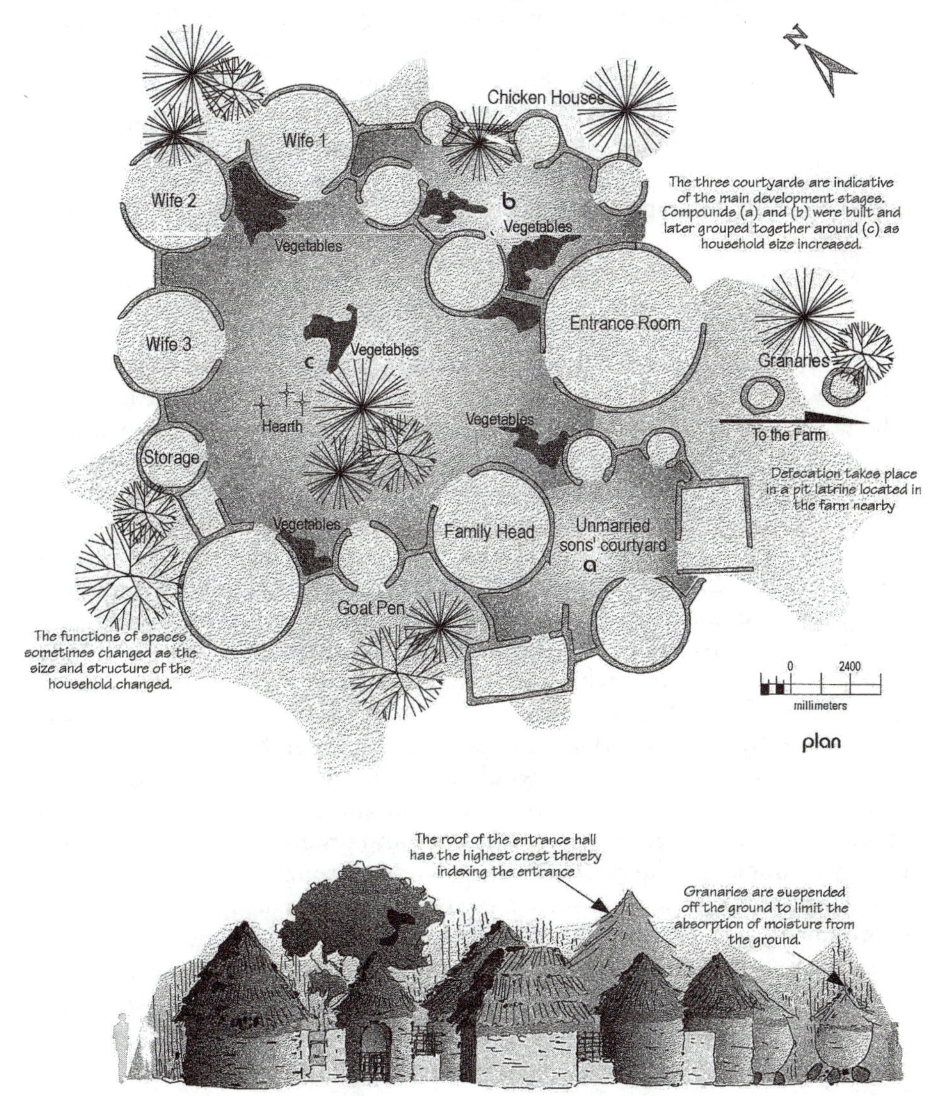

Fig. 7.1 A traditional compound household in Nigeria by Y Gbotosho

Yoruba house complex, drawn by Gbotosho, a Nigerian architect. Like the English country house, it separates male from female areas, but it is significant that unmarried women (like children) seem not to be worthy of designated space. In much of Africa girls come to womanhood through rituals of seclusion. Richards' (1982) detailed account of initiation rituals of Bemba girls in Zambia describes how girls are kept in a specially built hut for up to two weeks where older women use pottery figurines and accompanying songs to impart a combination of housewifely lore and teasing contempt for men. The activities of the initiation hut mark the movement of the girl from the free-ranging activities of a child to the more supervised and circumspect lifestyle of a married women. One figurine, which demonstrates the desire for autonomy curbed by a moral injunction to stay at home, is in the form of a hut with a swinging door; the verse that accompanies it goes:

Mwana alelila,	The child is crying,
Nshisalile uko alele	I did not see where it lay.
Tandabula.	The door is swinging.

Richards says of this: 'This song seemed to point to an important moral. Women told me that it was to teach the girl how to look after her baby. The child referred to in the song is crying because the door is swinging, i.e., it is left open because the mother has gone to a beer-drink and left her baby alone' (1982, p.208). This is certainly a powerful load of prohibition and desire to read into three short lines!

Bemba boys are not initiated and they are expected to roam freely beyond their villages, but we find that elsewhere in Zambia, for example among the Ndembu people, after the boys have been initiated by circumcision (which they say removes the female-like parts), they are exposed to forest lore and hunting; whilst Ndembu girls are brought to adulthood by being required to lie motionless, covered by a blanket under a tree which symbolizes the lineage. Both the Bemba and Ndembu have matrilineal kinship systems and, whilst it would be very wrong to suppose that there is any equation between matriliny and matriarchy, certainly there is no emphasis on patriarchal power. The woman is required to be passive and to stay at home for the benefit of her own kin group.

The sense of restriction I have been considering is also a theme in much of women's mythology and literature. One thinks of Cinderella in her kitchen, Rapunzel in her tower, Snow White in her glass casket. In these cases, all are virgins awaiting *release* through marriage, waiting to be made into women. But, when they are rescued, many women find nothing but another tower (see Fig. 7.2), as Irigaray (1992, p.47) wrote to her lover:

> You grant me space, you grant me my space. But in so doing you have always already taken me away from my expanding place. What you intend for me is the place which is appropriate for the need you have of me. What you reveal

Fig. 7.2 *Hawa Mahal*, The Wind Palace – a viewing structure, for court women, Jaipur, India

> to me is the place where you have positioned me, so that I remain available for your needs. Even if you evict me, I have to stay there so that you can be settled in your universe. ... You never meet me except as your creature – within the horizon of your world. Within the circle of your becoming.

Irigaray is not rejecting him, she does not know how to be without him, but also does not know how to be herself within the structures he has laid down for her. Later she refers to the way in which a structuring view which needs to appropriate and limit is the denial of possibility: 'Divided by lines, cuts, holes, walls ... life is not given in these or from these. It is severed from its becoming. From its perpetual renascence' (p.53).

Within the movement known as *l'écriture feminine* (see Box R) that Irigaray is part of, Cixous is particularly sensitive to the way in which constrictions of space and sensations of boundlessness can be employed as a means of communicating feeling between women with very different social experiences. In her novel *Angst* (1985), she strives to understand the power of love by trapping it in a stiflingly claustrophobic room of confinement, incarceration and entombment. A nameless dying man and a woman who loves him share a small, sparsely furnished hotel room with an attached bathroom. The experience of the death is written through the paralysing of the woman in space. The confined space becomes almost tangible, not as a place but as a force of restraint (see Box S).

BOX R *L'écriture feminine*

The emotionally charged way of writing about being female in a system of thought and language which is constructed in the masculine is a mark of the movement known as *l'écriture feminine*, associated with a group of French philosophers and psychoanalysts. They have often been accused of essentialism, but it is difficult to see how they could have taken on the project of writing their own being without it looking as if they were making statements about the condition of being female (in all situations). They deliberately strip away the social reference points like class, age, and nationality so that the 'I' in their writing can be perceived, free from the structures which they challenge, as they try to write themselves unmediated by oppositions to male-generated categories. It is not intended when they do this that they speak for all conditions of women, or that they suggest all women would be the same if they were deprived of their various social categorizations. The devices they use, however, can give the false impression that this is what they are doing. Of course, their task is impossible, but it is difficult to see how else non-oppositional difference between genders might be thought, how it might struggle into existence.

BOX S *Angst*

I had become, in front of him, passivity itself. I had become that formless space out of which I had to wrench each infinitely feeble impulse to think – an exhausting effort – in order to drag one single step forwards out of that nothingness. The door was quite near me, but somewhere else, and I couldn't reach it. Because I wasn't separated from it by space but by centuries of another time, I was trying to tear my voice free from terror, trying to shift the vast expanse of stagnant water. As if the distance between Him and what I was didn't exist – the distance I would never get over. ... To save myself I would have had to put space between the sentence and me, to protect myself from the shrapnel's deadly blast. I was caught, paralysed. ... My body was keeping me down: I couldn't even get to the head of the bed ... I was alone, not far from the bed, caught in the immense crystal of loneliness and I wouldn't survive. (Cixous, 1985, pp.199–200)

In this book time is suspended, characters are effaced, place is almost completely eradicated and space is condensed in order to render female emotion into writing. Cixous is attempting to transcend the boundaries of phallogocentrism through the inscription of the female body and, elsewhere, she quite specifically announces that she writes 'to get past the wall' (Cixous, 1991, p.5) and 'to make room for the wandering question that haunts my soul and hacks at my body: to give it a place and a time' (p.7). Here we have the familiar contradiction of desire to escape incarceration and longing for a place of one's own. She writes of 'boundaries and so many walls, and inside the walls more walls. Bastions in which ... I wake up condemned. Cities where I am isolated, quarantines, cages, "rest" homes ... my tombs, my corporeal dungeons, the earth abounds with places for my confinement' (p.3), but then sees 'continuity, abundance, drift' (p.57) as being especially feminine in writing.

At the end of *Angst* the man dies and the woman climbs up a tower, out of the confined space into radiant, unbounded sunlight, declaring, 'Suddenly no more steps, no more walls ... suddenly no more past', we might continue 'no more human relationship'. It sounds more like an afterlife than a life that she conjures into existence, a heaven which resolves the contradictions of structured existence by launching into the sublime, where the woman fuses into the man: 'I loved him as never before. Then I loved him: mad, unknown – gone' (Cixous, 1985, p.218). Cixous spatializes the female predicament and moves us far beyond a simple attack on patriarchy in her rejection of phallocentric law (Shurmer-Smith, 1994). For her, as for Irigaray, men are not enemies, but phallic power is and it can only be subverted through the relentless rejection of all structural oppositions. This necessarily involves the denial of spatial oppositions and boundaries, including notions of centrality and peripherality and it is as liberationary for men as it is for women.

L'écriture feminine is not to everyone's taste and it can certainly be very hard work to read, but, since it resides at the edge of experiment in generating new ways of thinking the world and abounds in spatial metaphors, it really cannot be ignored by cultural geographers. The difficulty emerges from its very resistance to logical categories and its requirement that every implement we use for thinking and feeling must be re-evaluated, but this is also its strength and emancipatory force.

Many of the French feminist philosophers draw upon the concept of *jouissance* as a feminine attribute with which to counter phallic power and its structures. There is no doubting that *jouissance* is very exciting to think, as well as to experience physically. *Jouissance* resides within the economy of desire/abundance, not the economy of want/scarcity. Whereas economics based upon scarcity concerns itself with controlling, conserving, excluding in order to maximize satisfactions, assuming (in the words of elementary textbooks) that 'man's wants are unlimited', economics of abundance prioritizes giving, enjoying and consuming. It works with words like *benefit*, rather than profit; it *evaluates* rather than counts and it is more impressed with *treasure* than with wealth. Before anyone says that this sort of idealistic thinking is fine for people who can afford it, but that scarcity will always dominate the thoughts and actions of most human beings, we would like to point to the truism that those with the least material goods often place an emphasis upon mutual support, whilst those with the most often seem fearful of losing them, are worried that their time will be wasted and are mean with their emotions. We insist that it is the interaction between people and things, not the amount or nature of the things, that determines the nature of economic thought, as is well demonstrated in the gift-exchange systems of non-market economies, where feasting and luxury are central to the notion of welfare.

Freud taught us to think of *economies* in contrast to *topographies* of desire, though his thinking was firmly grounded in ideas about *expenditure* and *investment* of energy. His metaphor emerged from neo-classical economic theory, though it was applied to sexuality, and it is little wonder that he could not work out what it was that women '*want*'. He saw women as a 'dark continent', but since in one society after another women are excluded from the control of goods, services and political power, even of their 'own' time and sexuality, we would suggest that they have learned to think and feel outside the legitimate structures based around maximization. Throughout the world, women generally don't *want*, they *need* or they *desire*, and this is why so many women's lives seem to be such a curious mixture of mundane practicality, with no room to manœuvre, and dreamy fantasy or petulant yearning.

Jouissance is the principle of benefit within the economy of desire. It is a term which does not translate well into English – 'enjoyment' is too half-hearted, but comes close; 'joy' catches its soaring or flowing quality but is not material enough. In French the word is used in a legal context to mean the enjoyment of rights in a property, what in English we call the 'usufruct' (the fruitfulness of the property, not the property itself), but it also means a sexual orgasm or any other sudden rush of pleasure or thrill. It is pretty obvious that this sort of benefit is not easily quantifiable or negotiable. It is, however, a route into conceptualizing a way of being in the world which confers autonomy without requiring that one becomes a 'master' oneself.

All of this has very real implications for how we may destabilize phallocentric power and, although superficially this may seem to be a concern 'only' of women, we insist that as we move from the structures of modernity to the fluidities (let alone flexibilities) of postmodernity the account-keeping of scientific logic seems a remarkably blunt instrument. In geographers' terms a shift from profit to *jouissance* means a cultural shift towards texture rather than dimension and an acknowledgement of boundlessness rather than a preoccupation with boundaries. Continuities, flows and sensitivities rather than categories, structures and measurements will inform a geography aware of *jouissance*. Whilst it may come hard for those with a vested interest in old constructions of knowledge to abandon the attempt to make it work, we can see from a great wealth of women's experience, as well as their writing, that there has been a long-standing scepticism regarding the scientific method and all its rituals. For a great many women it appeared that too many other things always had to be held constant for much scientific explanation of human behaviour to seem like anything other than a game with its own rules (even if it was a game which some of us learned to play quite well, though we were rarely chosen to be captain). *L'écriture feminine* subverts the rules of the game, primarily by overthrowing ideas about appropriateness and seemliness. It does this by using language which seems out of context, using domestic allusions in discourse on 'elevated' matters; referring intimately to the female body; not shrinking from love, angels, the soul, and generally embarrassing those who are accustomed to a less sensuous language. Acceptability in metaphor is a matter of power relations and we really need to think why, as Foucault (1984) remarked, geographical writing is so full of military and industrial language.

There are alternative views of the world, articulated, rather than constructed, by women worldwide and they frequently revolve around those things which educated men have focused upon as an indication of women's lack of logic – a preference for the particular over the general, for elaboration over the grand structure, for the nurturant, the sublime and the emotional. Doane (1987) has considered the importance of the melodramatic 'women's films' of the 1940s as a means whereby an alternative visual discourse emerged in which women were not cast in their usual film role as objects of male desire. But, as she shows, the films dwell upon persecution, hysteria, neurosis or paranoia, often related to:

> the deployment of space and the uncanny ... in which the house is foregrounded in relation to mechanisms of suspense which organise the gaze. This cycle of films might be labelled the 'paranoid women's films', the paranoia evinced in the formulaic repetition of a scenario in which the wife invariably believes that her husband is trying to murder her. (p.285)

She is referring particularly to such popular films as *Rebecca, Suspicion, Gaslight* and *The Secret Beyond the Door*. In *Rebecca* a young woman who has married a widower above her station finds herself constantly thwarted in her attempts to play the part of mistress of the house which her predecessor had made her own. She comes to believe that her husband had murdered his first wife and that she is herself at risk in what should be the safety of her home and marriage. The house becomes the site of her paranoia. Doane is not saying that married women are liable to be murdered, or that women are mentally unstable, she is suggesting that:

> A certain de-specularisation takes place in these films, a deflection of

> scopophiliac energy in other directions, away from the female body. The very process of seeing is now invested with fear, anxiety, horror, precisely because it is objectless, free floating ... to the extent that the cinema does involve an imaginary structuration of vision, it also activates the aggressive component of imaginary processes. (p.286)

What this amounts to is that in doubting or resisting structures generated around phallic power, a fluid normlessness may be conceptualized as madness or hysteria. Women of the war and postwar years flocked to see these films, finding in them an evocation of their own unease. They worked as myths, conceptualizing their own residual status, rather than as realistic representations of women's life.

As we have already shown in Chapter 4, there is a considerable genre of women's Utopian writing which deals with the tendency of women to think 'against the grain'. There is also a significant amount of women's travel writing from the West which allows white women to confront adventure outside their own societies in the presence of the Other. Yet the device is commonly used where there is room for only one heroine, moving, much like a man in a man's world, almost genderless, certainly sexless. Particularly good examples of this can be found in the work of Murphy (1965, p.xviii), setting off for the Orient on her bicycle or a mule:

> because in general the possibility of physical danger does not frighten me, courage is not required; when a man tries to rob or assault me or when I find myself, as darkness is falling, utterly exhausted and waist-deep in snow half way up a mountain pass, then I *am* afraid – but in such circumstances it is the instinct of self-preservation, rather than courage, that takes over. (Murphy, 1965, p.xviii)

For, whilst men's travel writing (at least as far back as Dumas) frequently stresses the sexual adventures to be had abroad, and where subaltern women are depicted as easily available, the women's equivalent emphasizes the possibility of adventure with only limited *fear* of sexual advances, which the heroine is more than capable of parrying. (Cesara's, 1982, autobiographical account of her field work in Africa is a notable exception to this and she was much criticized by the anthropological community for revealing her relationships with African officials. It is interesting that she published not only under a pseudonym, but also concealed the name of the country she worked in, so great was the transgression of the usual race and gender rules.) For all of its exoticism, as we have already shown, travel writing usually perpetuates structures 'at home' and rarely offers the chance of imaginative subversion, though it is certainly escapist (Bell *et al.*, 1993).

In the attempt to rethink women's being in space, there has been a relatively recent move to challenge some of the stereotypes about moral and physical dangers for women outside the home. We are familiar with men's glorification of the city as a place for the delight of the sensibilities of the gentleman *flâneur* through the work of De Certeau (1984),

Benjamin (1989) and Harvey (1989), but there has frequently been the assumption that such joys are not available to women, that women are threatened and vulnerable on the streets and are unable to gaze freely (Massey, 1991). However, Wilson's *The sphinx in the city* (1991) and Heron's collection *Streets of Desire* (1993) both seek to reappropriate the urban as a proper domain for women. They both recognize the freedoms women can experience in the anonymity of the large city, the release from petty surveillance associated with community living, the feeling of being responsible for one's own life, and the ability to move in the public domain. Though neither assumes that this is as easy for women as it is for men, both argue that in the contemporary city it is more possible for a woman to lead an existence predicated upon her own desires and will than it is elsewhere. Similarly, several of the contributors in Heron's collection *Truth, Dare or Promise* (1985) comment upon the sense of liberation they felt when they left homes in small towns or suburbs in search of their own identity. Walkerdine (1985, p.66) puts this sentiment particularly forcibly:

> I have a terrible fear of the suburbs: I just cannot bear the provinces, and especially the edges of conurbations. Just like home. The safe familiarity of the bay windows, the neat gardens, safety like a trap, ready to ensnare in its unfolding arms. The price, always having to live in London, the fear and lure of elsewhere, of always working, never stopping, for working was the way out, the only guarantee, the safeguard against the necessity to accept, to return, to give up. ... To stop is to turn, like Cinderella, from riches back to rags. But my home is paid for by my mental labour. What else, except the slippery slide-path back to being ordinary, at home with children in a suburb?

For others it is less intense, a matter of shops and coffee bars, the sense that one is at the centre of things, or that one can do mildly transgressive things, such as walking alone without comment. Wilson (1991, p.7) believes that the fear of the city is in fact masculine, but that it is projected onto women:

> the city, a place of growing threat and paranoia to men, might be a place of liberation for women. The city offers women freedom. After all, the city normalises the carnivalesque aspects of life. True, on the one hand it makes necessary routinised rituals of transportation and clock watching, factory discipline and timetables, but despite its crowds and the mass nature of its life, and despite its bureaucratic conformity, at every turn the city dweller is offered the opposite – pleasure, deviation, disruption. In this sense it would be possible to say that the male and female 'principles' war with each other at the very heart of city life. The city is 'masculine' in its triumphal scale, its towers and vistas and arid industrial regions; it is 'feminine' in its enclosing embrace, in its indeterminacy and labyrinthine uncentredness. We might even go so far as to say that urban life is actually based on this perpetual struggle between rigid, routine order and pleasurable anarchy, the male–female dichotomy.

Wilson's contempt is reserved for Utopian planning and utilitarian housing schemes into which urban households are decanted in an excess of zeal for rationality over stimulation. This she perceives as an attempt to impose order upon the city and deprive it of its feminine side. The safety, extolled by advocates of defensible space such as Newman (1973) and Coleman (1984) she sees as both an excess of surveillance, fostering conformity, and an excess of privacy and suspicion of outsiders.

Clearly, there is insufficient knowledge of what ordinary women really do think about with regard to the relative advantages of bland safety over excitement, and certainly women who write books about the city fall into a rather exceptional category about whom we can hardly generalize. We would simply like to draw attention to the fact that the whole of what is generally considered a feminine accommodation to space and place is frought with confusion – we simply do not know where women are.

8

Subaltern geographies

Emerging from Gramsci's work on hegemony (which assumed that the views of the dominant class would not simply be imposed upon the rest of society, but would be constructed as common sense and then more or less generally accepted) was the realization that in the experience of domination, subordinate groups of people generate identities which respond to their subordination. The term 'subaltern' was coined to refer to the people and cultures which existed, by definition, below the dominant; for Gramsci it meant, predominantly working-class cultures, forged in the light of their own lack of control, incorporating ideas and practices about resistance, but also those of compromise and acceptance. The condition of the subaltern is that of experiencing and living within the world as constructed by a dominant group, an experience which constructs one's own identity as an 'Other' to the dominant – one becomes by virtue of what one *is not*.

Today the term 'subaltern' generally refers to postcolonial situations, though technically it can be applied to any non-hegemonic position, as in the case of homosexual people in a heterosexually constructed society, women in conditions of patriarchy or phallocentrism, and children or the very old where they are debarred from full participation in decision-making. We can assume that in subaltern cultures not only are their people deprived, but their status of deprivation has become a part of the identity of the people. The condition of the subaltern is a product of a power relationship; it contains within it the idea of a view from *below* and emerges from a recognition of exploitation and conflict. Subaltern cultures are not original, preconquest, cultures fighting back against external domination, they are cultures which emerge within the context of domination and resistance, they have no being outside the asymmetrical contact with dominant categories. The subaltern exists as the 'Other' to the secure, establishment norm and, as a binary opposite, helps to define it logically, legitimizing its strength and 'normality'.

When one thinks of the way in which ordinary discourse in the West is constructed, this notion of subalternism snaps into sharp focus. It is all too easy to construct a series of 'Others' to one's own system of normality, Others that need explaining as deviations from one's own practice. Ideas about centres and peripheries, and concepts of remoteness or exoticism (not just where they are, but that they exist in one's thinking) have their roots in unexplored views of normality and (implicitly inferior) deviation from it. Such concepts are often unquestioningly incorporated into the researching and writing of geography as élite groups continue to replicate their view of the world as centred on, and thus explained through the structures generated by dominant groups in the West, however conceived. It is impossible to step far outside one's own experience, to see through the eyes of an Other, but one can be aware and critical of the constructions which arise from one's own standpoint (whatever that may be) and to question whether the construction is other than expedient. Many geographers and ethnographers have tried to study Others by looking through their eyes, but in doing so they are fooling themselves and their readers; one can engage, but one always remains oneself. Only the engagement can be represented or one falls into the trap of laying claim to the Other.

The film-maker Minh-ha (1989), wrote about the project of studying another people for its Otherness and rails against the unnecessary violence of a system of knowledge which presumes to define her very being:

> On one plane, we, I and he, may speak the same language and even act alike; yet on the other we stand miles apart, irreducibly foreign to each other. This is partly due to our distinct actualities and our definite history of involvement and power relationship. What I resent most, however, is not his inheritance of a power he so often disclaims, disengaging himself from a system he carries with him, but his ear, eye and pen, which record in his language while pretending to speak through mine, on my behalf. I thereby do not propose to eliminate, I'd rather make writing a site where opposites lose their essential differences and are restored to the void by their own interchangeability. Thus I see no interest in adopting a progression that systematically proceeds from generalities to specificities, from outlines to fillings, from diachronic to synchronic or vice versa. (p.18)

Minh-ha's project is to reclaim continuous, non-fragmentary ways of telling, to show that a story need not contain an oppositional structure, but may work with a suppleness. She aims to demonstrate that legalistic notions of relevance are unhelpful to understanding because they edit to the disadvantage of those whose meanings are polysemous, rather than singly constituted. The subaltern message is becoming increasingly familiar from postcolonial writers, but not from them alone (see Box T).

In the 1970s a group of Indian historians set about the task of reclaiming the subaltern histories of India, not just the history of Indians under the British Empire (though most existing written history had been constructed from the viewpoint of the colonial power), but also the histories of the 'silent' lower castes and classes in the colonial and postcolonial

BOX T *Cultural Studies*

Simon Armitage, in a poem called *Cultural Studies* thinks through his British working-class background in relation to the anthropology lecturer who seems to be condescending in her use of her experience:

> She would put down the myth
> of natural rhythm
>
> with reference
> to her cowrie-trading days
>
> in the black, African interior.
> How well she remembered
>
> their poor playing
> on her flageolet,
>
> and their indifferent footwork
> in the gentleman's excuse me.
> (Armitage, 1992, p.68)

This is a portrayal of an insensitive academic *mem-sahib* emasculating both the men of Africa and the man from the north; of course, she is represented as having no being or *jouissance* in her own right, a person existing on her memories of exotic empowerment.

setting. They were interested in the conditions which emerged in the light of domination. Their writings, in several volumes published in Delhi under the general title 'Subaltern studies', proved to be a focal point for the exploration, through film, poetry and the novel as well as academic study, from the point of view of those who were 'Other' to the dominant power. (Some of the most influential papers from the original books are grouped together as *Selected Subaltern Studies*, edited by Guha and Spivak, 1988.) They drew upon a range of Marxist and deconstructionist approaches to the condition of subordination and the mutedness and invisibility which accompanies it.

Unequal differentiation between categories of people is essential to the notion of the subaltern and this differentiation is often thoughtlessly incorporated into the 'of course' assumptions of hegemonic groups. If one has been socialized into a dominant group (as we suspect is the case for the majority of our readers) the subaltern view can be difficult to understand. All too often we find ourselves trying to 'speak for' subalterns, in our own language, or indulging in banal exercises of supposedly benign 'multiculturalism', celebrating superficial differences, rather then getting to grips with fundamental oppositions and, even more important, the structures which impose them.

In a brave paper entitled 'Roast beef and reggae music: the passing of whiteness', Jeater (1992) explores the problems of being white in a Britain which is racist. She writes of the difficulty of establishing an identity which, in a sense, *wishes* that it were not associated with the racism or with the history which underpins it. Recalling her rebellious adolescence in which she used reggae music in opposition to the rural middle-class 'respectability' in which she was reared, she shows that her political sympathies of anti-racism have no home in either blackness or whiteness in contemporary Britain. Although one empathizes with Jeater as a person, it is not difficult to see that her individual problem emerges from the logical opposition (which she seems not to reject) between black and white as categories within British society. It is not inevitable that those categories be generated or sustained and the personal anguish is best used as a starting point in trying to deconstruct them. Arguably the least of the problems of a racist society is that of the discomfort of individual members of the dominant category who would sooner not live with the guilt that accompanies their position. However, this unease is an instructive product of the structuring of society upon racist lines.

Spivak (1990), who is probably the most important deconstructionist theorist of postcolonialism, was born in India but received her higher education in the United States and teaches there. She recognizes the symptomatic significance of this experience of muting through assumed complicity:

> I will have in an undergraduate class, lets say, a young, white male student, politically correct, who will say: 'I am only a bourgeois white male. I can't speak'. In that situation – it's peculiar, because I am in the position of power and their teacher and, on the other hand I am not a bourgeois white male – I say to them, 'why not develop a certain degree of rage against the history that has written such an abject script for you that you are silenced?' Then you begin to investigate what it is that silences you, rather than take this very determinist position – since my skin colour is this, since my sex is this, I cannot speak. I call these things, as you know, somewhat dismissively, chromatism: basing everything on skin colour – 'I am white, I can't speak' – and genitalism: depending on what kind of genitals you have you can or cannot speak in certain situations. (p.62)

We have to admit that we usually feel a bit like the student in Spivak's class, our ideas about complicity, and our hidden feelings of easy complacency cornering us in a customary silence on issues of race and ethnicity. Terrified of being accused of liberalism, condescension, neo-colonialist attitudes or just simple insensitivity, we hedged uncomfortably about including this chapter at all, but decided that we were being cowardly. Spivak's comments came as a permission to enter, providing we 'develop a certain degree of rage against the history that has written such an abject script for (us)'. This is a rage which is not difficult to summon up. In large measure the rage is based upon the terrible *unnecessariness* of the categories which are used to construct groups which are mutual Others. In consequence, we agree with Bhabha (1988, p.11) that it is essential to acknowledge:

> the historical connectedness between the subject and object of critique so that there can be no simplistic, essentialist opposition between ideological miscognition and revolutionary truth. The progressive 'reading' is crucially determined by the adversarial or agonistic situation itself; it is effective because it uses the subversive, messy mask of camouflage and does not come like a pure avenging angel speaking the truth of a radical historicity and pure oppositionality.

Such that,

> our political referents and priorities – the people, the community, class struggle, anti-racism, gender difference, the assertion of an anti-imperialist, black or third perspective – are not 'there' in some primordial, naturalistic sense. Nor do they reflect a unitary or homogeneous political object. They make sense as they come to be constructed in the discourses of Marxism or the Third Cinema or whatever, whose objects of priority – class or sexuality or 'the new ethnicity' – are always in historical and philosophical tension, or cross reference with other objectives.

Bhabha is arguing that the ways of talking and writing about such issues as racism themselves are implicated in the thing which is being attacked, just as surely as a racist account itself is a part of, and not just a commentary on racism. This comes quite close to Spivak's rejection of the project of 'speaking for' – standing as the mouthpiece of a group, rather than speaking as oneself situated within it. Spivak's point is that the supposed insider who 'speaks for' a constructed group is as inauthentic as an obvious outsider advocate, in that he or she gains power from the generalizing role, a power which could not exist outside the tacit acceptance of the need to mediate. A person in an inferior position never *speaks for* someone in a dominant one. To be spoken for is inherently a mark of subaltern status; it also assumes a degree of homogeneity in those united in the representation of the speaker.

In a collection of South African poetry, published in 1968 at the height of *apartheid*, there was a liberal desire to deracialize which resulted in a classification of poets according to whether they originally wrote in English or had been translated out of Africaans or Bantu languages. This had the effect of hiding the very real discrepancies in life chances of the category of people known in South Africa as 'coloured', for these people might write in either English or Africaans and their names, too could be either English or Africaans. It is difficult to distinguish in the poetry a personal sense of alienation from a subaltern exclusion. The confusion has two effects, one is to reveal an essential similarity across race and this is something which many South Africans would have been unconscious of, but the other is to mask a constructed difference. This confusion was the very opposite of the everyday reality of the poets, for whom an unambiguous, legally-assigned racial classification would have controlled not only their political rights but also where they could live, how they would be educated, which medical services they could use, whom they could marry, which beaches they could swim from; a state of affairs concealing essential human similarity, daily emphasizing an artificially constructed difference.

For example, Jonker (1968, p.238), in her poem *I Don't Want Any More Visitors*, wrote:

> I want to be myself travelling with my loneliness
> like a walking stick
> and believe I'm still unique

The biographical notes at the end of the book tell us that the poet committed suicide, we can see the human anguish in the lines above, but we can't begin to construe it without knowing what categorization she was railing against in her cry that she wanted to believe she was unique. This sort of deliberate blindness to imposed, that is, politically determined, difference, though benign in its intention serves to deaden experience. In the Southern African politics of the 1960s, liberal whites, alternately 'speaking for' black and coloured people and denying that racial division 'mattered', were a brave exception to the norms of *apartheid*. It is now painful to acknowledge that the editors of the book, regardless of their good intentions, were replicating the power structure which constituted their everyday lives, whilst simultaneously presenting the richness of creative talent which cut across the divisions of society.

In an interview published in the journal *Radical Philosophy*, Spivak (1990, p.34) is explicit on the matter of the need not to obscure difference:

> People are similar not by virtue of being similar, but by virtue of producing a differential, or by virtue of thinking of themselves as other than a self-identical example of the species. It seems to me that the emancipatory project is more likely to succeed if one thinks of other people as being very different: ultimately perhaps absolutely different. On a very trivial level, people are different from the object of emancipatory benevolence.

It is important to remember that where the difference is structural it has been constructed in some interest, but we must also recognize that not all difference is structural; we may see this difference as being substantive.

The bulk of geography which takes any account of subaltern issues at all gets no further than 'speaking for'. Although there has been considerable soul-searching about the need to develop strategies of writing which give voice to non-élites and theoretical flirtation with the multivocality of the 'new' ethnography, there is still an exclusion of subaltern accents from academic discourse. The problem emerges from the entire academic project which decides what is 'appropriate' in terms of material, language and ways of theorizing and, unless the subaltern voice is 'framed' within a text constructed in the legitimized form (often literally by being 'boxed' or rendered into different typeface), it is likely that the academic profession will reject it as unsophisticated, or atheoretical, something which may be regarded as a source, but not as a contribution to debate. It all too often happens that people who have been used as valued 'informants' by researchers, and whose words have been used as the framework upon which monographs are constructed, are seen in a patronizing way when they presume to join in the discussion in their own right, using their own systems of knowledge.

At a conference on geography and music held in London in 1993 Rhys Mwyn, a practising rock musician whose group always sings in Welsh, delivered an address in which he denied that there was any nationalist element in the decision not to sing in English, saying, 'Welsh is the language I use when I go to the corner shop to buy my cornflakes. I'm not making any big nationalist statement'. He continued to say that he regarded himself as a culturalist, he did not want to see his Welsh culture, but neither did he want it frozen into the sort of unchanging traditionalism that would not allow the admission of global concerns or musical forms which were not generated in Wales in the distant past. He asserted that he believed in the 'celebration of difference', but then startled much of his audience by saying 'I'm not actually interested in Welsh identity ... I'm too busy to be spiritual'. He was greeted with a mixture of benign curiosity for the 'primitiveness' of his views, condescension for what was construed as a lack of political awareness, and outright opposition from those who thought that he *ought* to acknowledge that culture is always political. He was more glamorous, more attuned and more immediately involved in the subject of the conference than the academics he confronted, but they dealt with his different way of constructing knowledge by *marginalizing* it. The most often expressed sentiment was that of embarassment on the part of the academic élite who felt sympathy for someone who could not understand their project. This may seem like a trivial example, but the same process is at work when a mass of contributions from all sorts of Others are rejected by editors of geography journals who 'know' that they are basing their selection on their own absolute criteria of academic excellence or notions of suitability; it is also evident in people (including us) who write books and articles which hardly draw upon even academic literature published outside the USA and Europe. That the élite only attends to Others when they are playing the role of Other is one of the many double binds associated with the subaltern position. Spivak (1987) captures this thought beautifully in a paper describing her indignation at the way in which her contribution on the subject of her own outsider status at a conference was licenced and marginalized:

> In offering me their perplexity and chagrin, my colleagues on the panel were acting out the scenario of tokenism: you are as good as we are (I was less learned than most of them, but never mind), why do you insist on emphasizing your difference? *The putative centre welcomes selective inhabitants of the margin in order the better to exclude the margin.* And it is the centre that offers the official explanation: or, the centre is defined and reproduced by the explanation that it can express. (p.107, our emphasis)

She refers to the need to draw attention to marginality and the processes by which various people are marginalized:

> not to win the centre for ourselves. But to point at the irreducibility of the margin in all explanations. That would not merely reverse but displace the distinction between the margin and centre ... the deconstructivist can make use of herself (assuming that one is at one's own disposal) as a shuttle between the centre (inside) and margin (outside) and thus narrate a displacement. (p.107)

In this 'shuttling' she practises a form of situational selection, possible only to those who are relatively fluent in several modes of communication, but she is denying the 'speaking for' in favour of 'speaking as', even though she knows that she is stigmatized in doing so (see Box U).

BOX U From *Letter From Mama Dot*

The Black British poet D'Aguiar (1985) captures the cultural geography of being represented in his poem *Letter from Mama Dot*:

> You are a traveller to them.
> A West Indian working in England;
> A Friday, Tonto, or Punkawallah;
> Sponging off the state. Our languages
> Remain pidgin, like our dark, third,
> Underdeveloped, world. I mean their need
> To see our children cow-eyed, pot-bellied.
> Grouped or alone in photos and naked.
> The light darkened between their thighs.
> And charity's all they give: the cheque.
> Once in a blue moon (when guilt's
> A private monsoon), posted to a remote
> Part of the Planet they can't pronounce.
> They'd like to keep us there.

We find a variety of practical political and theoretical responses emerging from a recognition of the way in which contemporary society, and our thinking about it, is constructed out of difference. The most important of these terms of political movements are ethno-regionalism, racism and anti-racism, and 'identity' politics focusing on gender, sexuality and age. All of these, for purposes of expediency, are liable to invoke a measure of essentialism, homogenizing the individuals within their component groups, blurring internal differences in the name of political unity, constructing a sort of 'as if' identicality. The fact that we included racism on our list draws attention to the dangers of the question of identity politics if it is not linked to ideas about subaltern status, that is if it is not about exposing an asymmetrical difference. Preserving the 'purity' of one's own culture can rapidly turn very nasty indeed, as the worst excesses of fascism amply demonstrate. Notions of purity are inherently hierarchical, whether one is talking of the supposed ritual purity of the Brahmins within the Hindu caste system, of 'Aryan' purity in the context of Nazism, or of the purity of those of European descent which gave rise to the *apartheid* fiction that Africans were foreigners outside their assigned '*Bantustans*'. Purity is inherently dependent upon not just the presence of structuring but also its enshrinement in morality (Douglas, 1966, 1969).

One of the most notable features of the political scene in the latter part of the twentieth century has been the emergence of nationalist and ethno-regionalist movements throughout much of the world and, with this, a challenging of the construction of the unified states which we associate with nineteenth-century Western modernism. In the post-colonial era there has been an almost universal experience of the cry of the putative nations asserting their subordinate status within modern states. This has by no means been unique to those countries which were colonized by European powers, and the notion of 'internal colonialism' served to explain the desire for varying degrees of autonomy of nations and regions throughout Europe. (In the United States there have certainly been many ethnic movements, but they have not taken an ethno-regionalist or nationalist stance.)

Invariably, the political aspirations of ethno-regionalist groups are couched in cultural terms and frequently draw upon a cultural revival as a means of creating a sense of 'belonging', both in the construction of an homogeneity, based upon an assumption of a shared heritage, and in the marking out of a group as different from others, with whom they would otherwise be identified. An almost universal claim is that one's own culture has been lost or diluted by an imposed hegemonic power – a revival of what was one's own, in language, music, religion, cuisine, dress and 'custom', is common alongside a rediscovery and re-emphasis of history. Such revivals can take the form of 'nativism', the assumption that one is going back, insofar as it is possible, to conditions which existed in some 'golden age' prior to the loss of autonomy; or of renaissance, the birth of the culture in new form, keeping pace with changing conditions (for example, the Welsh rock musician that we referred to earlier). Obviously the two can exist alongside one another; a renaissance may well incorporate 'traditional' elements, but there can often be a tension between a conservative and a progressive view of ethnicity, a tension which does not necessarily mirror class or left/right ideology either.

Although it is inevitably presented the other way around by those involved, it is fairly safe to assert that groups of people who wish to blur their internal differences are responsible for the revival or creation of ethnicity, rather than that old cultures spontaneously reassert themselves and pull together those yearning for their lost past. It doesn't feel this way as one basks in the warmth of enthusiasm generated by, say, a Runrig concert, waving one's arms in the air chanting the word 'Alba', the ancient name of Scotland, with several thousand similarly moved people; it *feels* like plugging into a deep history. But it is a new Scottishness which looks to Europe, which rejects its old relationship with the British Empire, which unites the Gallic-speaking Highlands with the Scots-speaking Lowlands, which is likely to be socialist in its politics and youthful in its stance. It may not be the Scotland of Bonnie Prince Charlie, Walter Scott or Adam Smith, but this Scotland is also alive and kicking, and one whose progress was charted by Trevor-Roper (1983). But there is no reason to associate ethnic fervour with depth of pedigree. As Spivak (1990, p.34) asserts: 'I'm much more interested in the enabling violation of the post colonial situation than in finding some sort of national identity untouched by the vicissitudes of history'. This does not mean that isolating particular practices and claiming them as one's heritage is not an effective way of creating a valuable sense of identity; indeed people have often been prepared to sacrifice everything in the name of the struggle for their ethnic solidarity. The contemporary Sikh desire for Khalistan has been translated into a militancy which draws upon a history of struggle against Muslim invaders in the Punjab, but the opposition now is to Hindu India, which ancient Sikh heroes could easily have accommodated (see Fig. 8.1). It has been suggested that much of the current Sikh struggle draws upon a nostalgic longing for home on the part of highly educated Sikhs in Canada, the USA and Britain (Appadurai, 1990). Certainly a great deal of the wealth of the Punjab comes from the invested savings of expatriate Sikhs, and it is difficult to say whether one is looking at change within a traditional framework or as a part of a transformation. Either way, a consequence of the change in the Punjab has been an espousal of a fundamentalist version of the Sikh religion and a drawing apart of Hindu and Sikh Punjabis which is now almost as marked as the earlier separation of Sikh and Hindu jointly from Islam. The former easy intermarriage of Sikh and Hindu is a thing of the past; the practice described by Pettigrew (1975) of Hindu Punjabi families giving one son into the Sikh religion is now unthinkable, as is

Fig. 8.1 Sikh Golden Temple at Amritsar, India

the account given of a Punjabi village in the early 1970s, where the distinction between Sikh and Hindu was seen as religious rather than ethnic (Hershman, 1981). History may explain many aspects of Sikh identity, but bloody opposition to Hindu India is certainly not one of them.

Meanwhile, a perhaps trivial Punjabi cultural trait has swept across northern India – it is becoming increasingly rare to see educated middle-aged or young urban women wearing saris other than on formal occasions, and not invariably even for these. The Punjabi *salwar-kamiz* is becoming ubiquitous and is subject to seasonal fashion changes. It is no longer an emblem of Punjabi, and particularly Sikh culture – it is part of being a modern Indian woman and is seen as infinitely preferable to modernizing via 'Western' fashion. At a time when there is a fear of Sikh terrorism, it is perhaps strange to see women appropriating the freedom of Punjabi dress in a superficial denial of the levering apart of the people of northern India.

McDonald's (1986a) work on constructions of contemporary Breton identity picks up this same idea of women's ambivalence towards cultural signifiers as they work out a subaltern position which fuses issues of gender, class and sophistication with ethnicity. Rural women of Finistère had come to believe that it was in the best interests of their own and their families' social standing if they could combine the roles of hardworking peasant wives with the refinements of urban 'ladies'. This meant speaking French in public, wearing fashionable clothes and serving French cuisine. Doing the best by one's children was perceived as involving speaking to them in French, rather than Breton, a language which women came to reserve for their husbands, elderly relatives and animals. An urban-based, middle-class, ethnic revival in Brittany drew nostalgically upon the peasant lifestyle of the Breton far west and women found themselves curiously caught up in their aspirations towards sophistication, discovering that what they had been encouraged to reject was now valued not just as ethnic authenticity but also as a vehicle for the working out of women's issues, the nationalist and the feminist movements being intimately intertwined in Brittany. Yet the women who were expected to embody the Breton ideal and nurture the 'mother tongue' had espoused a version of modern living which was anti-nationalist, anti-ruralist and anti-feminist and they found it difficult to understand the romantic view of Celtic culture held by their university-educated offspring. It was even more difficult for them to understand the resentment of their children who blamed them for not having taught them Breton, when the mothers had believed (correctly) that French language, education and manners were the passport into the very middle class which was now seeking to capitalize on its difference. The women's liberation is sought as an emblem of 'the' Breton cause, but they find that, 'The movement has a discourse which, no doubt well intentionedly, invokes local women, but which these women are, in any case, powerless to contest' (McDonald, 1986a, p.182).

This movement goes well beyond Brittany and its relationship with France, taking in all the perceived Celtic oppression aspiring to the construction of a pan-Celtic identity through recourse to the romantic history we referred to in Chapter 4. But, as McDonald (1986b, p.333) emphasizes, there is no independent pre-existing ethnic group of Celts: 'The case of the celt is only one exemplification of a more general process of definition and self definition, in which categories of identity are constructed and come alive, not in isolation or in nature or in the mists before time, but in specific, changing, contexts'.

The subaltern status is, in part, constructed at the same time as ethnic identity; here a myth of oppression had to develop in tandem with a political movement which was not entirely concerned with injustice but cut across a spectrum of class positions. The status of the peasant proved useful in unifying the Celtic past and romantic views of peasant life, like Hélias' *Le Cheval d'orgeil* (1975; filmed in 1979) emerged as part of a nativistic process at the same time as the growing sense of Celtic identity. We can observe the use of Celtic themes in Scottish and Welsh nationalisms, where deprivation, marginality and cultural distinctiveness are spun into a common thread of mournful longing and militant defiance. Often folk-rock musicians like Gilles Servat and Tri Yann in Brittany, Runrig and Battlefield Band in Scotland and The Dubliners in Ireland, all with a pan-Celtic following, are socialist in politics and draw upon similar themes of exile, oppression and resilience, whilst employing traditional and contemporary musical forms and instruments in a powerful construction of subaltern consciousness.

Not all ethnic subalterns have regionalist or nationalist aspirations. Cultures of conflictive contact exist in a multitude of different situations, predicated upon such diverse bases as colonial expropriation of authochthonous populations, slavery, migrations of rural people to urban areas, counterurbanization, international migration (whether as temporary workers, settlers or refugees), retirement and second-home purchase (whether abroad or in 'marginal' areas of the 'home' country). Some of these result in nothing more than a rather banal marking of difference – the distinction between people from the south-east of England retiring in Cornwall is very slight, and though on occasions indigenous Cornish will stress the 'foreignness' of the incomers, more frequently they seek to submerge it (Plester, 1993). This said, there is often a profound marking of difference translated into a desire to control one's 'own' space and institutions to create places in one's 'own' image. Jackson's *Maps of Meaning* (1989) is particularly useful in its accounts of the metaphorical mapping of difference and deprivation onto the urban environment. He charts a subtle re-ordering of urban lifeways which rarely goes so far as to result in bounded ghettoes, but where networks of ethnic similarity are thrown across a city, overlapping with other networks. The point, in a multi-ethnic society, is that the maps are constituted out of areas of contestation, rather than out of absolute territoriality; one group's world is superimposed on several others', sometimes in overt friction, other times in differences in ways of living, they then become genuine maps of *meaning*, not just of districts with social identifiers.

Any subaltern group is likely to find itself in ambiguous spaces, zones of transition, not just in the usual geographical sense but also in the presentation of symbolic transitions between the pure and the impure, the safe and the dangerous, the known and the unknown. A poem by Alvi (1993) catches the sense of being an Indian in London:

> I live in one city.
> but then it becomes another.
> The point where they mesh –
> I call it mine.
>
> Dacoits creep from caves
> in the banks of the Indus
>
> One of them is displaced.
> From Trafalgar Square....
>
> In the double city the beggar's cry
> travels from one region to the next.
>
> Under sapphire skies
> or muscular clouds
> there are fluid streets
> and solid streets.
> On some it is safe to walk

Although this is most obviously imagined in terms of urban ghettos, *bidonvilles* and slums, alienating housing estates or immigrant quarters with exotic shops and restaurants, we can just as easily think through visible and invisible gay and lesbian spaces of safety and expression as outlined by Valentine (1993), Moos (1989) and Jackson (1989).

Perhaps the most overt spatial expression of subaltern status is to be found within the geography of the Hindu caste system in rural India, where the most stigmatized groups live outside the rest of the community and their section of the village is felt to carry a miasma which is polluting to members of 'clean' castes. There are many sociological accounts of this, but a compelling example comes not from an academic source, but from a short story by Devi (1993) about a woman from the cast of Doms, the 'outcastes' responsible for the disposal of the dead, who live removed from others who fear them. The woman, whose task it is to bury dead children becomes grief stricken by the work when her own child is born. Accused of being a witch, she is then thrown out by her own community to live alone on the other side of the railway tracks, in the multiple stigma of outcaste, poor, woman, and witch. The story is written by a woman from one of the highest Hindu castes who is exploring her own relatively élite subaltern status by imagining and claiming affinity with the condition which would take her to the very lowest point within her society.

Subaltern cultures require a consciousness that the intellectual projects of the hegemon cannot unlock one's own sense of being; it assumes that being outside is one's daily experience. Said (1993) shows how, once one is aware of the subaltern status, even the classics of Western high culture become imbued with the taint of exploitation, when he gives a reading of Jane Austen which accepts and normalizes the slave trade as the basis of genteel English country life. It becomes the task of the subaltern artist and intellectual not simply to oppose but also to deconstruct hegemonic representations, reading for overt repressiveness and for omissions, too. Subaltern intellectuals have often been very receptive to poststructuralist thinking, finding in it a means whereby one can escape the trap of always having to fight on the terms of those who have superior power; for with deconstruction it is possible to oppose whilst rejecting the categories which define one in the position of opponent. For this reason, there is such interest in the concept of hybridity, the possibility of society which can contain within it non-dialectical difference. Hybrid cultures are much more than syncretisms, they are not mixtures from two or more sources, but a creation of something new out of difference (see Box V). There is no reason, however, to assume that this necessarily means benign cultures of tolerance of the mulitculturalist variety. Hybridity should mean no more than that the various ways of being and thinking available are continuous, recognizing that segmentation is contingent and that ruptures are willed. It is in the alliance of subaltern groups that the logic of the core will be defeated and classification at the margin will become meaningless.

BOX V Hybridity

From Nichols' *Wherever I hang* (1989, p.10):

> And so I sending home photos of myself
> Among de pigons and de snow
> And is so I warding off de cold
> And is so, little by little
> I begin to change my calypso ways
> Never visiting nobody
> Before giving them clear warning
> And waiting me turn in queue

9

Outside the middle class

There has been considerable debate as to whether the class structures of Western societies are fragmenting and realigning or transforming into newer structures based upon a postindustrial economy or regime of flexible accumulation and disorganized capitalism (Lash and Urry, 1987; Harvey, 1989). There have been further debates as to whether class, with its basis in productive position, is being replaced altogether by a new stratification system based upon the ability to consume, which produces consumption cleavages or sector alignments. Clearly, as we pointed out in the introduction to this section of the book, society is also fragmented along other lines of force, such as age, race, ethnicity, gender and sexuality, but there has been a resurgence of interest in systems of stratification in Western societies, particularly where restructuring has produced an *underclass* (a term first used by the economist Myrdal in 1962) of individuals who are separated off culturally, socially, politically and economically, well below what have become known as normal living standards; people who neither produce nor consume successfully.

If we glance back in history, it seems there has always been some kind of stigmatized underclass in capitalist societies. For example, Marx noted the existence of a category below the working classes which he called the *Lumpenproletariat*, or residuum of unemployables. Marx and Engels saw these individuals as a politically insignificant group, a judgement that persists today, despite evidence to the contrary (Heath, 1992). Engels (1892, 1952, p.60) described this residuum as 'the lowest stage of humanity', and their living conditions as consisting of 'masses of refuse, offal, and sickening filth ... the atmosphere is poisoned'. Some hundred years later Harrison's (1985, p.208) description of an inner-city area of London is strikingly similar: '... smells, refuse blowing everywhere, vermin, frequent arson ... and possibly diseases.'

Superficially it might seem that there has been no transformation at the bottom of our class structure. It would seem that there has always been a conception of a polluted, brutalized, potentially criminal underclass in Western society. However, in describing those in poverty in such a way, both Engel's and Harrison's well-intentioned descriptions stereotype and stigmatize individuals by ascribing a sense of *undesired differentness* to them (Goffman, 1963; Gans, 1993) and in doing so, they deploy stereotypes that highlight aspects of pollution. Douglas (1966, p.5) highlighted the power of ideas of cleanliness and contagion as a means of maintaining distinct segments in rigidly structured societies: 'It may seem that in a culture which is richly organised by ideas of contagion and purification the individual is in the grip of iron-hard categories of thought which are heavily safeguarded by rules of avoidance and by punishments'.

When we categorize individuals along our lines of power and desire, we use terms relating to pollution, dirt and disease to symbolize a distinction between order and disorder, and to establish boundaries within societies, but we also use notions of pollution to legitimize actions that would otherwise be morally repugnant or simply patronizing. So, for example, it becomes morally justifiable to talk of helping those within the underclass to get out of their filth or to get *their* act together.

By both the UK's relative and the USA's absolute measurements there does seem to be a growing number of disadvantaged individuals. Empirically, social inequalities are increasing, firstly due to local experience of global economic restructuring which results in flexible urban poverty and, secondly, because of a related restructuring of welfare payment systems. The increases also reflect the device of conceptually assigning many new transgressive groups to the category of the underclass. Indeed, many have argued that behavioural deviance and disadvantage seem to go together in a vicious circle to form an underclass. The anthropologist, Lewis (1966), was one of the first to argue this in his (in)famous culture of poverty thesis. Based largely upon his research in Mexico, Lewis argued that the urban families he was living with were generally apathetic and provincial in outlook, and lacking in a sense of community organization. He linked these characteristics to the high birth and divorce rates, a confusion of sexual identification and an inability to plan for the future. He argued that these characteristics were largely self-perpetuating between generations as individuals were ill-prepared to take advantage of changing social, economic and political conditions.

A number of criticisms have been levelled at Lewis's work. Firstly, for a model of human behaviour, it is clearly far too static for people do move in and out of poverty between generations. Secondly, he misses the positive aspects of what we might term a rebellious counterculture. Lewis is generalizing a theory of cultural poverty from the analysis of only a few *lifestyles* of poverty and it is doubtful whether a homogeneous culture of poverty really exists as described by him (Valentine, 1968). On inspection his arguments seem to be more a construction of middle-class fears about an apathetic yet dangerous underclass, threatening their own values and structures, and we come back to Douglas's ideas of pollution and categorization. However, Lewis's work remains an extremely powerful thesis politically and academically; it has been widely used to blame the poor and deny them both citizenship and benefits on the grounds of their lack of responsibility within their own neighbourhoods. Lewis's arguments have been taken up and disseminated, particularly through the work of Murray (1990), but as Mann (1992, p.2) argues, the major problem with all studies of poverty is that 'it is all too easy to slide from identification of a social group who suffer social problems into the position where the victims are regarded *as* the social problem'.

Lewis's arguments have been challenged by the work of the American sociologist, Wilson (1987, 1989), who has emphasized the spatial expression of poverty in the idea of the ghetto and processes of hyperghettoization. He outlined an ecological model of the formation of ghettos of poverty in cities in an attempt to escape the negativity of Lewis's work. He argues that socially and racially segregated areas of Chicago have emerged due to processes of spatial, economic and industrial restructuring (Wacquant and Wilson, 1989). Such ghettoes – or inner-city areas of entrapment – are seen to be perpetuated by systematic outside discrimination. When discrimination lessens, as it did in the 1970s for blacks in the USA, leaders of the community move out, leaving behind the hopeless underclass. Paradoxically therefore, the social movement of some individuals leads to social entrapment of others as poverty breeds spatial and racial exclusion in urban contexts.

The problem with Wilson's arguments is that they all too often fall back into the behavioural version of poverty expounded by Lewis and Murray, of a criminal underclass rationally maximizing their wants via their state benefits (Bagguley and Mann, 1992). Fainstein (1993) has argued that paradoxically Wilson's work confuses and de-emphasizes the racial problems of inner-city areas. Moreover, Wilson's ecological explanation also fails to explain the complexity of factors involved in any sociocultural movement, such as the generation of place myths and the use of key metaphors in defining turf for identity (Shields, 1991b; Marcuse, 1993). Both Lewis and Wilson's work as well as the term 'underclass', have served to obscure the analysis of social processes instead of revealing them (Bagguley and Mann, 1992; Gans, 1993; Keith and Cross, 1993; Morris, 1993; Wright, 1993). It is too great a synthesizing term for such a diverse range of individuals (Gans, 1990). There is no inherent condition of being poor and evidence suggests that in their culture and aspirations, the poor are not so different from those in either working-class or middle-class positions (Heath, 1992; Willets, 1992, Wolch *et al.*, 1993).

Part of the obfuscation produced by the term 'underclass' is due to the fact that as we move into a postindustrial economy, where three million unemployed is fast approaching the norm in Britain, there is increasing confusion between what we feel to be our respective needs, wants and desires. At the simple level of commodities, what constitutes a luxury and what constitutes a necessity is increasingly difficult to specify. It has long been accepted that every generation seems to reinvent the idea of poverty by transforming some of their predecessors' luxuries into necessities and upgrading the material conditions of what is still poverty by virtue of the lack of ability to consume effectively within the conventional terms of the society. So, for example, we find that attitudes to central heating and car-ownership shift, both because of changed provisions and changed expectations.

For members of the middle class, the confusion and uncertainty in the prioritization of needs, wants and desires helps to produce the range of lifestyles that are often taken for granted. But this same sense of uncertainty also exists for those on low wages, the dole or social security. Individuals on unemployment benefit, or even on student grants or loans do sometimes spend some of their income on a range of desirable commodities like clothes rather than saving for when the heating bill arrives, but one definition of poverty is the inability to exercise choice about the disposal of one's income. The consumption of status-rich commodities can be primarily associated with desires to identify with and participate in particular subcultures (Hebdige, 1979). Alternatively, those whom we commonly associate with the underclass may try to aspire to the notions of respectability of their former lifestyles, in an attempt to preserve their public dignity and their social obligations; this can result in going short on 'necessities'. Parts of the underclass are very much like the mainstream, in privileging their desires over their needs at times. This very fact has led to both academics and politicians claiming the moral high ground, patronizing those in poverty by assuming the they should always prioritize their needs over their desires, adding to the debate as to what should be done to alleviate and eliminate extreme poverty, almost invariably conceived of in terms of material need. Yet we would argue that it is desires that make social life significant. (If we go back to the story of Cinderella, we could easily position her within the underclass, but if she had not prioritized her desires over her wants and needs she would never have had the chance to marry Prince Charming!)

We can see that our understanding of poverty as a relative concept must take account of the intersection of lifestyle, gender, stigma and place. We can begin to see how much of the Western food aid to countries in Africa is *mis*placed in that it relies too heavily on absolute judgements of what famine is from a Western viewpoint, as opposed to the socio-cultural processes of how it develops and can be coped with within the country itself (McCann, 1986). We can also see that images of dire destitution (Fig. 9.1) serve to construct notions of relative affluence and therefore lack of entitlement amongst the poor.

In Western countries similar symbolic processes are apparent on the streets, in homelessness. Homeless people are stigmatized as the most disadvantaged part of the underclass in Britain and the United States

Fig. 9.1 Mother and child

because they lack proper shelter. In most social policy accounts, the 'home' in homelessness tends to be simply equated with housing – homelessness simply becomes rooflessness (Somerville, 1992). However, issues of homelessness go much deeper than simple definitions of shelter which may be central to Western social and cultural attitudes. Wolch and Dear argue, in their book *Landscapes of Despair* (1987), that homelessness is a symbol of that part of the deinstitutionalization process which failed. (Deinstitutionalization referring to the policy of releasing ex-offenders and the mentally ill, to be cared for in the community.) They estimate that half of the homeless people in the United States are victims of economic circumstances and gentrification processes (see Burt, 1992; Goetz, 1992; Gans, 1993; Hebdige, 1993), a quarter are ex-psychiatric patients (see Koegel, 1992) or drug users, and a quarter are victims of personal crisis or natural disaster. The problem of homelessness lies not in the victims themselves but in the wider context of their stigmatization in both the academic and popular press (Wright, 1993). In addition, the stigmatization is emphatically gendered for the homeless, and women outside the structure of home and family are often depicted as crazy, immoral, feckless or supernatural (Rowe and Wolch, 1990), whereas men are seen as simply down on their luck.

Homeless individuals are often portrayed as extremely mobile, tramps, nomads, drifters and transients by choice or temperament. However, many homeless people simply have to keep constantly on the move because of harassment by police, criminals and 'respectable' locals symbolically defending their patch from assumed pollution. For example, in 1991 a police operation, significantly code-named *Cinderella*, to clear the streets of litter and crime, pushed the homeless off many of the more sheltered streets of Manchester and into suburbs and other cities. The police argued that this was because they were a health hazard, but of course, there was also the city's image in relation to its Olympic bid to think of. In London tube posters now ask people not to give to beggars but to a formal charity (see Fig. 9.2). Under Indira Ghandi's state of emer-

Fig. 9.2 Underground poster, London, London Transport Museum

gency in 1976 squatter settlements and urban slums were similarly razed to the ground, and there is the continuing scandal of the 'tidying away' of street children in Rio. Clearly we come back to notions of pollution and disorder versus social control when the destitute are removed from view so blatantly.

Spatial movement for the homeless is, however, generally restricted due to the fact that they have to carry all their possessions with them, using carts, crates of supermarket trolleys (Smith, 1993). Any voluntary movement over long distances is related to offers of social support and like the rest of society, most homeless people are stayers, rather than movers (Rahimian *et al.*, 1992). As Wolch *et al.* (1993, p.159) argue, 'homeless people may move around for the same reasons that homed people do – to meet needs [wants and desires] for food, shelter, income, friendship, and various services'. A major problem is that the media either shows the homeless as a public nuisance, or drowns them 'in such lugubrious sentimentality that they are rendered helpless puppets, the pathetic other, excused from active civic responsibility and denied personhood' (Smith, 1993, p.89).

There have been some notable attempts to transcend such stereotypes, by enabling the homeless to re-enter the domestic economy. *The Big Issue* is a magazine that was set up in London in 1991 to help the homeless by getting them to help themselves, but the very ethos which underpins it is based upon an enterprise-oriented notion of charity, with overtones of Victorian values which link entitlement to work. The paper is sold by homeless and ex-homeless people who keep 60 per cent of the cover price (the rest goes on production costs). At first sight the magazine looks like a viable alternative to begging, but it has been the centre of considerable debate since some argue that it is now impossible to move around central London without a copy for fear of being hassled by sellers, and that it is only 'a mixture of guilt, compassion and persuasive sales pitches' that accounts for its booming circulation. Others argue that it gives the homeless a voice (*The Big Issue* has recently published a book of poems and photographs by its vendors called *From a Sheltered Flame*), and dilutes the unpleasant relationship between beggars and the public. (Whether this dilution should take place is itself a matter of controversy, given that real contradictions in society are being concealed in the act of creating disguised unemployment.) Clearly, editorial content of *The Big Issue* is a form of aestheticization of poverty (Harvey, 1989) and may help to propagate the very stereotypes it is seeking to undermine, but it also serves as a process of empowerment in its parody of other forms of journalism. However, given that the magazine is a site of advertising and is underwritten by liberal business interests, we may also see it as an interesting ideological and symbolic mixture in its commodification of both charity and poverty.

In the United States we can find similar arguments centred around

Wodiczko's *Homeless Vehicle Project* (Hebdige, 1993; Smith, 1993). At once a parody of professional designer merchandise and a potential home for homeless individuals, the project carried all the meanings associated with the German world *unheimlich*. Hebdige (1993, p.180) describes the 1988 prototypes as consisting 'of a hinged metal unit which could be extended to provide sleeping, washing and toilet facilities as well as a can storage compartment. The product has been tested by a panel of homeless consultants and adapted to the precise subsistence needs of its prospective users'. Hebdige argues that we should see the *Homeless Vehicle Project*, and films such as *The Fisher King* (Gilliam, 1989), as tools for examining social relations in the city, provoking and identifying conflicts, tensions and alliances between our ideas of home and homelessness, rather than as simply artistic representations. The very fact that contemplating, representing and catering to dire poverty can become a niched activity is one more fold in the flexibility of accumulation.

For many homeless individuals it is often the loss of a valued home and family that gives rise to their outsiderness. Rather than admit dependency in the simple shelter of a hostel, many prefer to live on the streets, whilst continuing their search for a mythical home. Our normative definitions of home discredit the personal worlds of those in poverty, who may often be contesting their categorization through in-between concepts like un-home (Veness, 1993).

Homelessness is clearly intended to be the other side of the 'home idea' (Veness, 1992), or 'homeyness' (McCracken, 1989), that we outlined in Chapter 3 and showed to be both ambiguous and open to manipulation. But reflecting the polysemy of home, there are conflicting ideas of what homelessness means to different individuals. Many women admit that they have no home of their own but deny that they are homeless; others have a home and claim to be homeless (Somerville, 1992). There always seems to be an ambivalence in our relations with both ideas of home and homelessness. The mythical attraction of the home with its plethora of significations is often a site of entrapment and normalization of cultural values for both women and men (McDowell, 1983; Saunders, 1989). Where this is rejected the myth of freedom of the homeless takes on a romantic flavour, close to ideas about the sacred liminality of poverty outlined by Turner (1974, p.232):

> I would like to point out the bond that exists between communitas, liminality and lowermost status. It is often believed that the lowest castes and classes in stratified societies exhibit the greatest immediacy and involuntariness of behaviour. This may or may not be empirically true, but it is at any rate a persistent belief held, perhaps more firmly by the occupants of positions in the middle rungs of structure on whom pressures to conformity are the greatest, and who secretly envy, even while they morally reprobate those groups and classes less normatively inhibited.

The theme of homelessness is pervasive in American literature. As part of the American pioneer dream the physical and spiritual sides of home and homelessness are by no means easily separated. We can see this particularly in the writings of Henry Miller (1956), which maintain that it is more manly to be destitute having had a good time, than to accept a meagre rationalized existence year in, year out. In Jack Kerouac's *Lonesome Traveller* (1962), we find an account of the American *hobo* or homeless brother. Kerouac brilliantly captures the ambivalence with which our society treats those who are homeless: 'In America camping is considered a healthy sport for Boy Scouts but a crime for mature men who have made it their vocation – Poverty is considered a virtue among the monks of civilised nations – in America you spend a night in the calaboose if youre [*sic.*] caught short without your vagrancy change' (p.149). Here the contradiction plays upon conflicting notions of freedom and rights.

In much of women's literature we find a longing to escape the confines of the home by living on water, for example on a houseboat or a yacht; to be homeless in relation to structure and law, yet at home in relation to oneself. For example, in both Duras's novel *The Sailor from Gibraltar* (1985), and Nin's *The Four-Chambered Heart* (1950), we find a sense of transgression and movement in living in a space beyond the bounds of society. In both we find the convenient conflation of being at home and drifting which forms much of the cultural representation of gypsies. We also find that amongst the themes of exoticism, mysticism, drugs and love, of alternative hippie and New Age lifestyles, there is a desire to live in the cracks between the structures of the traditional Western home-based society. Whilst such groups may aestheticize poverty, they also explore much wider societal metaphors about drifting in and out of liminal realms.

But behind all the hippie talk of communitas in the 1960s and 70s there also lay a stark individualism, a cultivation of the Self and the American dream of the frontier (Hall, 1968). The hippie era was about translating middle-class values into some sort of rebellious protest, mostly stylistically through language and clothes. Hall (1968, p.4) described the hippies' linguistic style as 'an emphasis on the continuous present tense – "grooving", "balling", "mind-blowing", "where its at" and their prepositional flavour – "turn-*on*", "drop-*out*", "freak-*out*", "be-*in*". "love-*in*". "Cop-*out*", "put-*on*", "trip-*out*", "up-*tight*".' Clothes tended to symbolize the assumedly primitive way of life of the American Indian, or the spirituality of India – moccassins, bells, beads and headbands – in a commodified appropriation of the cultural Other, but some very old fantasies based around the interplay of structure and communitas were being worked out at the same time.

More recently, in Britain, the phrase 'New Age travellers' has become associated with a diverse range of individuals whose lifestyles transgress but do not subvert the conventional middle-class values associated with the home as a fixed abode. As Hetherington (1991) argues, many travellers are there partly because of the notoriety and glamour attached to the lifestyle, and their identities are formed around the hostility shown towards them. They display their lifestyles publicly through the consumption of signifying commodities, 'associated with satiating bodily desires, through food, alcohol, drugs and ritual adornment' (Hetherington, 1991). More often than not, this *visible* bodily transgression leads to contests over the meaning of particular sites. In England, Glastonbury has long been regarded as one of the most carnivalesque places for a diverse range of students and travellers who claim the area is enshrouded in myths of King Arthur and the Isle of Avalon. Its commodification and development into a much promoted, highly priced and policed annual festival has created a drift of more 'purist' groups to more obscure locations for their celebrations. For many it has become routinized, no longer a free and liminal experience but a highly controlled mediatized event and the tricks of those who get in free are publicized and applauded as honourable transgressions of an unwarranted takeover.

In Britain, the contested meaning of certain key places for those living outside or on the edges of the middle class has been channelled into a series of increasingly acrimonious environmental protests. For example, in 1993, at Twyford Down, near Winchester, England, a carnivalesque environmental protest against the construction of a motorway, involving 500 dancing and largely peaceful individuals, focusing around New Age 'tribes', quickly turned into a conflict when some 200 private security guards attempted to forcibly breakup the action. In the United States, Smith (1993), has shown how, in 1991, Tompkins Square Park in Manhattan similarly served as a site of contested meaning. Used by homeless people to build shacks which not only provided shelter but also drew attention to poverty in the midst of affluence, the clearing of the park became a focus for different notions of citizenship and entitlement.

Despite the aestheticization and commodification of notions of poverty, homelessness and Otherness, these transgressions should not be dismissed lightly for they contest the assumptions, generally taken for granted, that those more or less inside the comfortable structures of 'homeyness' place upon those outside. They expose the compromises which have been made to achieve security and can cause unease in those who have taken the life course which implies acceptance of the line of rigid segmentation. Moreover, whilst some representations can be viewed as simply romanticizing poverty, they perhaps also explore an archetypal, Western myth of boundless space beyond family and home, structure and law, a rhizomic line of flight which is dangerously seductive and which threatens the edifices of ordered society.

10

As old as you feel?

This chapter begins with a cliché, and like all clichés, it is an easy thought about a potentially difficult issue. Geographers, like other social scientists have frequently made use of the concept of age (remember the old joke about mapping populations broken down by age and sex?) and on the face of it, one would think that age was one of the few social categories that does not need to be agonized over – age is just a matter of quantification, of course; a simple objective fact with no room for subjectivity. We are increasingly familiar with the knowledge that cultural constructions of gender are not the same as biological definitions of sex, but there is little popular questioning of the cultural biology of age – everybody knows, of course, that 'the young' are exciting and impetuous and 'the old' are conservative and doddery, and that this is a natural consequence of their physical and mental stamina. But, like many of the things people say 'of course' about, there is no 'of course' at all. Age is as culturally constructed and personally negotiated as any other attribute and we don't just mean lying about one's age, whether it is to get into bars 'under age' or to make oneself eligible for a job which has an upper age limit.

We think it is important to think why age *limits* exist in the first place. Given that age, as opposed to physical capability, has no necessary characteristics, such attributes as age has are granted culturally. Societies which take a chronological view of the life course mathematize the typical human career, bringing it under control through measurement and bureaucratization. But there is no necessary reason why lives should be looked at in this way, why they should be seen like tape measures being played out, a single strand, carefully calibrated. Indeed the idea of keeping a precise note of age is a relatively recent one even in Western societies and the general marking of birthdays seems only to have become common after the introduction of compulsory schooling. In her contemplation of working-class childhood, Steedman (1989) draws upon the writing of Henry Mayhew, an early social commentator examining the conditions of the poor in the East End of London. In the winter of 1849/50 he encountered a little girl, making her living by selling bundles of watercress, who overturned his notions of childhood with her composure and self-reliance. She certainly was not supervised by any adult in her commercial transactions as she moved from the wholesale market to the places where she sold her watercress, and he reported her as saying 'I ain't a child, and I shan't be a woman till I'm twenty, but I'm past eight. I am'. Bewildered by the categories she was using to think her position Mayhew noted. 'I did not know how to talk with her'. Western adults often have similar reactions to pictures of children working (see Fig. 10.1).

Fig. 10.1 A child metalworker in Jaipur, India

In highly structured societies people often find it difficult to think difference without more or less rigid categorizations separated by boundaries. With age categorization, time becomes a flow that is constantly metered and regulated and the problem arises that one becomes more concerned with measurement than with substance. We have already referred to Deleuze and Guatarri's differentiation between the smooth and the striated, whether in space, fabric or thought, and would argue that ways of approaching age can be contemplated according to whether a lifetime is perceived as smooth or striated, a continuity or a set of separable categories. The difference has considerable implications for geographical distributions of people at the local and regional scale, since the categorizing view is likely to result in separate provisions for what are deemed to be different sets of people. So, for example, it is easy to assume that an ageing population represents a problem of *care* and requires provision of special *sheltered* accommodation for *elderly* people (Corden, 1992), conflating a language of protection and dependency with that of age. Whilst many old people do require care, the conflation works to restrict and demoralize those who do not.

Social scientists are interested in age because it is used as a basis for generalizing about people's attributes and behaviour. Like any other classification system it is assumed that with some degree of confidence one can 'read off' likely characteristics and attitudes from knowledge about age and that the more classification systems one can superimpose on the same population, the more accuracy one will have in one's predictions. So most of us would have a more convincing stereotype of a person if we superimposed the categories female, middle-aged, married, white, professional and ex-colonial. It is precisely for this reason that questionnaires, whether generated by governments, social scientists or market researchers, rarely avoid a question about age, whatever topic they are investigating. But we are all aware that not everyone who is sorted into the same 'box' is the same, or makes the same use of the attributes conventionally handed out (so, for example, the description above didn't prepare you for the fact that a person sorted by those categories might like mountain bikes and rock music). The very fact that we are sometimes surprised by people's behaviour demonstrates that we do have expectations about the categories we generate.

So the first problem we have to confront when looking at age is that of generating the categories. If we use terms like 'young', 'old', or 'middle-aged' we are obviously working with a blunt instrument. At what point does 'young' become 'middle-aged'? Do people aged six months have much in common with people of 15 years? You can see straightaway that the two questions are of a different nature – the problem with the first is that we need to be able to blur the line between the two categories, the problem with the second is that we need more divisions. We are thinking about transitions from one category to another in a different way at different points in the life of the person. It is here that age is very distinct from any of the other sorting principles, for people never stay still within their category – each day you are much the same age as the day before, but it clear that you are on a moving path.

Different societies have a variety of ways of making sense of the ageing process and making the categories they generate seem as if they flowed naturally and unavoidably from the accompanying biological changes. We find that everywhere the issue of age is intimately tied up with moral concerns – questions of what *ought* to be the attributes and behaviours, rights and obligations, associated with particular points in life. But the morality associated with age, particularly in Western societies, is a curious thing; it is assumed that it keeps in check what is unacceptably animal in humans, but it is always represented as being based on natural goodness rather than mere expediency. Morality is an instrument of social control and we both become very suspicious when people talk about it.

Most people in Western societies have become familiar with the bureaucratic management of age. It seems too obvious to be noteworthy that legally we have to be registered within a specified number of days of our birth, and that from this date of birth we count in years. We accept that certain things legally must be done or may be done after particular birthdays. For example, in Britain one must follow an approved course of education after the age of five and may not stop doing so until the age of 16; one may marry with parental permission and be heterosexually active at 16; apply for a driving licence at 17; vote, sign legal documents, and marry without parents' approval at 18; but not (male) homosexually active until 18; claim a state old-age pension at 60 or 65 (depending on sex); and one must reapply annually for one's driving licence after 70. Apart from all these legal matters, there are the local, personal, class and gender assumptions about what is 'proper' at different ages – when one is old enough for a first bike, to date, wear make-up, have sex, marry, reproduce, get a job, or leave home; also when one is considered too old to stay out all night, dance, wear short skirts, be sexually active, or know one's own mind. There will be conventions about when one should defer to one's elders, be old enough to know better, be mature enough to shoulder one's responsibilities, or be too old to be expected to help. All of these tend to be regulated through such things as encouragement, reputation, gossip, ridicule, anger, ostracism, but they are subject to negotiation individually and over time. They also configure differently over space and clearly it is this package of laws and conventions which the geographer is trying to catch when age is mapped. The age structure of a place will make a difference to its character, though what proportion of the population is working, what sort of housing, welfare, transport and leisure facilities are to be found, how many educational institutions there are, all depend not on age as an absolute, but on the way in which age is culturally constructed in the particular local situation.

Although societies with bureaucratically managed age do have 'milestone' birthdays, like 18 as the age of majority, whereby one is a child one day and adult the next, we more usually think in terms of age as a flow without major breaks. There is a certain anxiety about the decadal birthdays as personal reminders of life-cycle changes, but there is something of a tendency to treat them defiantly and to say that there is no special significance to be attached to any particular point in a numerical system. This is, however, not reflected in the way in which social scientists collect statistics, 10-year tranches are frequently constructed for purposes of generalization. Local common sense tells us that at some points in our life flow the divisions of attributes based on age are very close together, at others they are spaced far apart. Some of this is to do with the rapidity of physical growth and degeneration, but much is to do with the speed at which one is allowed to change one's behaviour.

The notion of the life cycle is closely allied to that of age, but it really should not be equated with it. Many ideas about age-appropriate behaviour are to do with the way in which particular groups of people try to routinize their domestic cycle, particularly the behaviours attaching to sexuality, child-bearing and rearing, household management and care of the elderly. The way that mothers of young children behave, becomes the expectation of the 'normal' behaviour of women who are at the age when they 'normally' have young children in that society; women who have their children earlier or later than 'normal' will find themselves acting in a remarkably similar way to other mothers 'growing up quickly' or 'being kept young' – and are likely to find that their friends come from the age group of the 'normal' mothers, simply because of their shared identity and interests (Monk and Katz, 1993). In recognition of the particular way in which the mothering role shapes women's lives, Tivers' (1985) work on the mobility problems of mothers of young children was structured by the age of the children, not the age of the women.

When thinking about the domestic cycle, we are more interested in generation than in age, as such, and most of us know from bitter experience that intergenerational relationships are notoriously fraught with power struggles and conflict. Does one argue with one's father because he is 36 years older and he has the behaviour specific to that particular age, because he represents ideas that were laid down before one was born, and he was young during another historical era, or because he is one's father and there is a structural opposition? The answer is probably all of these but in different mixtures at different times and in different places. It is important, though, to appreciate that the apparent age attributes of individuals are constructed out of their interaction with other people of the same and different ages, genealogical positions, histories and biographies.

In many societies economic, political and kinship positions are intimately related. In rural India, for example, a young man's 'boss' is likely to be his father, a young woman's is probably her mother-in-law, and the 'head of household' will often represent the rest of the group to the outside world. Ideas about how much autonomy young people can have and how much respect middle-aged people can command are very much influenced by situations like this. It is common to find that men in societies with a peasant-based economy are kept in a 'boy's' status for far longer than we would expect in urban industrial societies where they go out to earn a 'man's wage'. This phenomenon was first revealed in Arensberg and Kimball's (1940) classic study of a peasant in County Clare, Ireland. In one peasant society after another one finds that men in their thirties continue to ask parental permission to do things that urban adolescents would take for granted: they often have no money they can call their own and are known in the community as their father's son, not themselves. Often they marry late. Not only are ideas about paternal and filial duty different, but so are notions of maturity and morality, for to take on a full adult male role 'too soon' would be to challenge one's father, to go on acting 'young for one's years' would be to undermine the gravity of the role of pater familias and the standing of one's family. When peasantry is overlain with tourism or with labour migration, there can be traumatic changes to age statuses as young people begin to accumulate wealth in their own right.

Traditionally many subsistence societies had some form of age-setting which served to build a degree of social homogeneity into groups of people of roughly the same age, but, perhaps more importantly, also made for absolute discontinuities in the flow which modern societies imagine age to be. We can particularly associate age-setting with the pastoralists of East Africa, such as the Nandi or Masai. The negotiated aspect of age is not to be found in such situations, for ambiguity is ironed out by initiating groups of people into adulthood over a specific period of time and conferring a common social age upon them. The groups of coevals are then known as age sets. People coming to maturity after a set is closed will then wait several years until the next one is opened and will be given the same social age as those initiated in the next batch. These sets pass through various age grades, which map out the life career and have particular activities and privileges attached to them (for example, warrior, householder, ritual expert). Life, then, was not conceived of as a continuous slope but a flight of steps, where absolute divisions are more or less arbitrarily imposed. Although we have spoken of people in general, it is very rare to find the age-setting of women; the divisions are usually associated with age grades based upon masculine roles and adult women are seen as a continuity, where the homogeneity is disturbed, not by formal passage of groups through publicly oriented age grades, but by the developmental cycle of the domestic group. A woman may become a mother, a mother-in-law, a grandmother or the mother of men in particular grades, but her changing status will be derived from others. (This is symbolically marked by the fact that in East and Central Africa, the majority of women are named teckonymically, that is as 'Mother-of-senior son'.) This way of conceptualizing age has local implications since it is often associated with the assumption that young men will be excluded from the main residential community, in cattle-herding, education or warfare, and that young women will be closeted within it, losing the freedom to roam which is granted to children of both sexes.

Much of Chapter 6 was concerned with the problems of coming to terms with the transition from boyhood to manhood and revolved around the emotional difficulties associated with the recognition that, in becoming a man, a boy becomes 'the same' as his father, who now has to become different because he is no longer the father of a child. We saw how ideas about threat and abdication were worked out through notions of nature and culture, with the wilderness, wildness, and sport playing an important part in marking the liminal stage through which boys are passing. In less dramatic form, we saw in Chapter 7 how adult women are socially created through the enclosure of adolescent girls and their separation from open space. It is quite common that, as with the Hima of Uganda (Elam, 1973), this maturing takes place as the first stage of a marriage ritual and girls who do not marry never achieve adult status.

The removal of younger men from the domestic realm parallels a common assumption that they are in a liminal phase, not quite in society. Frequently there is an expectation of unruly behaviour which invokes expressions of horror in their elders, but which also serves to create a buffer between the obedience of childhood and the relative autonomy of adulthood. This liminality is frequently associated with the wilderness. Some of the best known examples of this come from East Africa where the wild behaviour of young warrior age grades is legendary. Clearly this stage of life is used to establish codes of masculinity based around strength and aggressiveness, and a separation from the settled world identified with women and children. For women it is much rarer to interpose a lengthy liminal stage and maturing takes the form of a reigning-in of girls' spontaneity. Their liminality will usually be marked with incarceration, restriction of mobility and bodily restraint, and marks a passage from the status of child to adult without a lengthy period during which youth roles are assumed. Whereas the young men's liminal behaviour is an ideal type of masculinity, the women's is a void between two contrasting statuses.

Katz (1993) has outlined how in Sudan rural children are allowed considerable licence to move around the village lands, in complete contrast to urban American (middle-class) children who are seen to be vulnerable to criminals and traffic. But she also notes that as Sudanese girls reach puberty they are confined to a narrow domestic realm. Children of both sexes support the seclusion of the adult women of their household, since their home range is wide enough for them to run the errands and carry the messages which allow the women's range to be so very restricted. Whilst American women are accorded more rights to move in the public world than are children, they often find these severely curtailed by their own sense of danger or anxiety attaching to certain kinds of public domain as well as by other people's ideas about respectability (Valentine, 1993).

Schildkrout's (1978) study of Hausa children in Kano, Nigeria, asks the radical question, 'what would happen to the adult world if there were no children?' Here she is not thinking about a replacement generation, but, '[I]n what ways are adults dependent upon children? What is the significance of children in maintaining the relative status of men and women? Why do people want or need children?' (p.111). For Schildkrout, children were not just the means whereby urban women's purdah could be supported, they also served as a conceptual category of asexual freedom in a society where gender-marking was absolute for adults. Children could trade in the markets in collaboration with their home-based mothers or on their own accounts, they could roam freely in and out of other people's households (whereas adults could not) and became a source of information and gossip. Schildkrout concludes that 'it is important to take a perspective in which one views children and adults as complementary participants in the social system' (p.133), a view which is not at all common in modernist social structures where children are seen as minors, learners and dependents.

Age-based conflict is built into the very physical fabric of some societies. In Tanzania the Nyakyusa people live in villages where all the men will be of roughly the same age. It is considered important that as boys reach puberty they should move out of the parental village, setting up their own villages with boys of a similar age from the same neighbourhood. The boys will herd their fathers' cattle, bringing together beasts from several adult villages, and they will be fed in groups by their various mothers until they acquire wives of their own, whom they will bring into their village. When their sons reach puberty they, in turn, will move on, but their daughters will live with them until they marry. Men, however, will go on marrying as many wives as they can afford, always drawing upon the same age group of unmarried women and, as a consequence, fathers and sons are potentially in competition for the same young women. Since men inherit their fathers' wives, as well as cattle and political status, it has been assumed that the reason for the removal of young men from the villages of their fathers is to minimize political, economic and sexual jealousy between them. It would be dangerous to read the Oedipus complex onto a social order for which it was not generated, let us simply say that the complex addresses the same sorts of father/son conflict in a different context. Certainly the Nyakyusa resolution of the tension has interesting spatial outcomes. It is, however, important to emphasize that the rigid marking of age is, once again predicated upon the need to differentiate between categories of men; women are regarded as more or less homogeneous and will be spread fairly evenly by age through the men's age villages.

The construction of the category 'young' in opposition to 'mature' is one which is easily taken for granted and throughout Western society there are a host of ways in which the category 'youth' is set apart as surely as it is in rural Tanzania. Whether we are looking at street culture, youth groups or boarding schools, there are spatial separations of young people from the mass of adult society. In the collection *Dare, Truth or Promise* (1985), edited by Heron, on British girls growing up in the 1950s, one account after another shows the girls who are today's middle-aged women, learning to separate themselves from their homes and their parents in bids for personal autonomy via education with its consequent upward social mobility. Valerie Walkerdine, is particularly moving on the way in which it was necessary to lose the status of treasured daughter in order to strike out on her own and avoid the suburban destiny which was otherwise hers. The association between adult personal autonomy in London versus claustrophobic domesticity a Yorkshire suburban setting was, for her, absolute and linked to the need to keep working. The suburb as central, the city as escape – the view of a peripheralised person, seeking a line of flight through a dream: 'You should never have educated us, the ordinary girls of the fifties. We are set on becoming, and you will not stop us now ... leaving all the pain and uncertainty behind in that other place' (p.76). The only escape involved a physical move and a class treachery which accompanied a movement from childhood to adulthood at a time of rapid social change and opportunity.

We may contrast this example with that provided by the anthropologist, Turnbull (1985), who presents his experiences at the major English public school, Westminster, in order to make artful comparisons with adolescence among the Mbuti pygmies. And yet we find the same separation from family, the same attention to tiny details of personal choice and self-assertion that are the means whereby a separate identity is constructed. Turnbull was right to see parallels with the overtly different Mbuti case because, whereas Walkerdine used her youthful separation to construct a line of flight, he, like his Mbuti coevals, was given space to work out his own places in a structure he had little desire to escape. His account is paralleled in thousands of factual and fictitious school stories, normalizing the liminal stage between privileged boyhood and élite manhood, stories of schools which served to construct adults in the approved image of the society.

Recent work on Caribbean societies demonstrates that transition through the stages of life does not have to be formalized and suggests that where men are not powerful heads of domestic groups, competition between males of different generations diminishes and relations between people of different ages become more relaxed. Pulsipher's (1993) study of domestic groups on Montserrat shows that not only do related women form the core of domestic groups, but that men attach to these groups in non-exploitative ways, whether as brothers, sons or lovers; marriage is seen as a bid for a certain kind of respectability in later years, rather than the institution upon which the family is based. Pulsipher shows that women and men of all ages can have remarkably equal relationships with each other and that, because marriage is not structurally important, intimate friendships between people of radically different ages can easily be integrated into gender relations. It becomes interesting to think about the possibility that age structures are socially significant only where there are patriarchal relations.

In support of this, we find that the management of age takes on a different complexion in the move from industrial to postindustrial societies and we notice that the deconstruction of age categories is a characteristic of postmodern culture, where more individualistic relations are the norm (see Box W). Featherstone and Hepworth (1991) are particularly interesting on the way in which markers of age are becoming blurred for relatively affluent members of Western societies as the attri-

BOX W Postmodernity

Whilst there has been much debate as to postmodernism (see Box K), there has been rather more associated with the notion of postmodernity, seen as an epochal shift out of the structures of modernity. An evident slippage in the use of postmodernism and postmodernity has produced a widespread confusion. And, with individuals seeking to date the arrival of postmodernity, the transition between modernity and postmodernity has been considerably oversimplified. Habermas (1987) has been particularly critical of any discourse of postmodernity, arguing that the overarching rationality that characterized modernity still exists in Western societies. Nevertheless, there is considerable evidence for a recent shift in many Western societies away from rigid structures based, in part, upon industrial manufacturing or Fordist economic systems, towards different structures predicated, in part, on flexible, post-Fordist, service-based economic systems that privilege individual desire for those who have access to money (Lash and Urry, 1987). However, we should not characterize this as a wholesale shift, the experience of modernity and postmodernity are in no way antithetical to one another.

butes of 'youth culture' are appropriated by both the young and the middle-aged (whilst the end of the middle-age keeps receding, eroding all but 'deep' old age). We notice that the clothing, sports and cultural preferences of this section of society tend not to be heavily age-typed – for example, similar jeans and casual wear in general are worn by toddlers and their grandparents and both could find themselves negotiating the nursery slopes at ski school.

The media has played an important part in this age deconstruction as the very young and the very old are exposed to the same cultural images, including topics which in different eras would have been used as a means of creating separate age identities by exclusion on the grounds of unsuitability. The old conspiracy, which pretended that children knew nothing about sex and the elderly had forgotten, has lost its power in the face of shared film and television viewing; one only has to watch the film *Forever Young*, for example. People of different age groups now receive the same information about current affairs, nature, the arts, and they are exposed to the same humour, advertising and sports coverage. To some extent they develop a common discourse, just as they would have done in the premodern era before the rigid construction of childhood and when old age was not a lengthy stage of life. In the modernist era people in Western societies absorbed a notion of childhood which assumed long dependency and considerable vulnerability. Many are shocked to discover that this was a relatively recent social construction. An exhibition at the Bibliotheque Nationale in Paris in 1993 focused on the way in which the lengthy modern childhood can result in a stunting of talent which would formerly have been allowed to flourish (Sacquin and Ladurie, 1993). We can interpret this as a consequence of a process of structuring for industrial purposes, a rationalizing division of labour which resulted in the establishment of the patriarchal nuclear family associated with modernism. As new family forms emerge, along with new concepts of the person, we can see the increasing precociousness of children, which is often interpreted as pathological by those whose values were laid down in the era of high modernism. But we would not want to exaggerate the contemporary collapsing of age categories. People of different ages may have more in common with one another within postmodern culture than they have had since before the industrial revolution, but there is still a great deal of age-based prejudice in both directions and this has very clear geographical outcomes as the attributed or chosen lifestyles of particular age-groups come to be configured in space.

The most obvious manifestation of this is in the phenomenon of retirement migration whereby older people move in search of improved environments, often in lower-cost settings than they had previously lived. Part of the counterurbanization in Europe is based upon retired people moving out of urban areas, or people nearing retirement making a move in anticipation of ceasing work. In France this commonly takes the form of a move back to rural areas from which one's family originated, where there are ties which have been maintained over a lifetime of vacationing amongst other urban people who regarded the region as 'home' (Shurmer-Smith, 1991). In Britain, however, people commonly move to areas where they have little direct experience. The British practice means a heavy concentration of retired people in seaside resorts and rural settings, often attracted by purpose-built accommodation in the form of bungalows or flats.

The American phenomenon of the village entirely devoted to 'senior citizens' has yet to arrive in Europe, but settlements such as the 'Sun City' described by Fitzgerald (1986) can only be explained in terms of a deep ideological age segregation. These villages are built speculatively and properties sold or let only to retired people, usually above the age of 50, but the irony is that people are attracted to them by the promise of a lifestyle which denies the stereotype of advancing age as a time of inactivity. Such settlements commonly focus on a golf course and country club, they stress leisure rather than simple lack of work and it is expected that their residents will participate in a great many sports, clubs, classes and entertainments. They are places where older people, outside the disapproving gaze of the young, are able to pursue their own interests without being ridiculed for still wanting to dance or wear fashionable clothes. Yet there is something defeatist about such environments – the young are excluded, other than as visitors or service workers, and the settlements represent a retreat from an ageist society rather than a challenge to it, perpetuating the age structures rather than subverting or transgressing them. As Fitzgerald remarks, this results in a paradoxical impression of 'childishness' in these obsessively neat, conservative, organized communities.

The residents seem to be playing at youth as it is formulated in the same structures that would consign them to a stigmatized old age; as such they are a far cry from the denial of the social and emotional value of negative stereotyping we find in those who have genuinely refused age. Writers as far apart as Sand (1929) and Duras (1990), both in their seventies, have written about the way in which much of the frailty of old age is imposed by those who have not yet achieved it. Sand never gave up writing, not only to support herself but also an extended family, and maintained an active interest in the theatre and politics; Duras continues to write prolifically, works in film and theatre and delights in perplexing the world with her intimate relations. Neither withdrew into a ghetto in order to remain active, and both seem remarkably close to the Caribbean women described by Pulsipher, independent householders with an identity they derive from no one but themselves. The diaries of Nin (1974) similarly reveal a woman uninterested in her age as a categorizing principle. She remained convinced (and convincing) of her physical appeal until her death in her seventies and had no sense of the inappropriateness of this. This is very different from the angry but resigned attitude to ageing that we find in de Beauvoir (1972), which accepts the validity of the category 'old'.

In a similar vein, my fieldwork on Ile-aux-Moines in southern Brittany has shown an affluent island community, predominantly made up of retired professional people where, though there is no emphasis on busy-work and energetic recreation, there is a reluctance to be excluded from active life (Shurmer-Smith, 1992). The island is a treasured heritage and its descendants do everything in their power to obtain property there for holidays and for retirement, though it offers few opportunities for employment and people leave to make their careers in Paris. The island serves as a 'summer place', as young people descend upon their elderly relatives who provide not only food and accommodation but also boats to sail and contacts with young people attached to other households. People of all ages are likely to attend the same social events and be involved in the same gossip networks. Like so many holiday resorts, the island offers a privileged environment for fairly wealthy retired people, but, unlike Sun City, it does not construct a ghetto of elderly fun which is visited only as a matter of duty and from which residents cannot extend hospitality.

The modern Western construction of age categories to create artificial behavioural differences results in the mapping of people into age-appropriate environments which imply a disabling overprotection of the children and old people whose family resources can afford it and simultaneously, a neglect and marginalization of others. In spatial terms this means a restriction to a 'safe' home environment and encapsulated long-distance travel on supervised tours and trips. It is little wonder that much fantasy writing constructs ideal worlds which run counter to this, in order to offer an escape from the claustrophobic cosseting. So, for example, C.S. Lewis's *Narnia* (1964) stories have children magically wandering for years unsupervised in terrifying and wonderful circumstances, whilst, simultaneously, resting at one instant in time, in the bounded protected space of a wardrobe at home. Lessing's (1979) Canopean Empire is dominated by people who have overcome 'the Degenerative Disease' we know as ageing, and those aged over 200 are still active and attractive. We also find women in Gilman's *Herland* (1979) not experiencing a menopause but continuing to reproduce and be politically important. Time travel is another trope whereby historical change can be imaginatively freed from physical degeneration and it is important to see this as a metaphor for the denial of the need for social age categories. So in *Back to the Future* films, for example, we find the protagonist at school with his mother or holding his great-grandfather on his knee, as age is disconnected from time. Where writers are experimenting with new worlds they do not just invent new political and economic orders, they rethink age and gender, two forms of structuring which are intimately related.

The Utopian retirement village advertising itself as 'The town too busy to retire' (Fitzgerald, 1986, p.203) can be seen as playing upon the same fantasy of agelessness by theming active retirement. By excluding children and providing no facilities for those in 'deep' (i.e. inactive) old age, its residents free themselves from the oppositional categories which would otherwise prevent them from claiming 'youth' but, of course, they cannot win; if there are no old, there can be no young.

Age is a myth for structuring both thought and individuals more generally. Deleuze and Guattari (1987) draw upon *The Crack-Up*, a short story by Fitzgerald (1936), to elaborate the ways of thinking about the life process and the person. They take Fitzgerald's story of cynical withdrawal from romantic enthusiasm into selfish bitterness as a device for thinking about different types of change which happen not just to the body but also the way of living with the self (though they question whether the route has to be the same as the one his character takes). Fitzgerald's story is one of increasing density, moving from a creative writers' youthful assumption that the world is comprehensible and the hero is in control, through a depressed acceptance of relative failure, to a late middle-aged abandonment of the initial project of greatness and its replacement with a knowledge that selfish success is achievable. Fitzgerald writes of the early construction of the ageing process as being one of easily recognizable stages, assuming that 'Up to forty-nine it'll be alright ... I can count on that', but then the realization, in his thirties, that physically he has been imperceptibly cracking all along and that youth has gone without a clear division being made. Fitzgerald depicts the last point as a 'break' with the former order – 'the ones who ... survived had made some sort of clean break. This is a big word and is no parallel to jailbreak when one is probably headed for a new jail' (1965, p.52).

Deleuze and Guattari (1987, p.200) take the story as a way of understanding their distinction between 'the line of rigid segmentarity with molar breaks' (the heavily structured system of categorization), 'the line of supple segmentation with molecular cracks' (the flexible structures which we can associate particularly with service-class postmodernism), and 'the line of flight or rupture, abstract, deadly and alive, nonsegmentary' (the rejection of structure, whether rigid or supple, the step outside). They summarize this as 'Break line, crack line, rupture line' (p.200). (Fitzgerald's 'break' is their 'rupture'.) We may see the structured ageing strategies of East Africa as dominated by break lines, certain and unavoidable; those of the eternally active Sun City residents recognize crack lines, insidious in their encroachment. The strategies of Duras, Nin and Sand, with their outrageousness and refusal of categorization, are rupture lines, seeming wilful and perhaps uncomfortable to those onlookers playing more rule-bound games.

Neither Deleuze and Guattari nor Fitzgerald is constructing a homily for gracious ageing. They are using the changes in body, mind and spirit which are wrought by age as a way of thinking about a view of life and personhood assuming that it is only with the rupture entailed in the line of flight that one can achieve the sublime. 'Now one is no more than an abstract line, like an arrow crossing the void. Absolute deterritorialization. ... One has painted the world on oneself, not oneself on the world' (pp.199–200). Age was an obvious way to think this adventure of flight, whereas the animated exclusion of the disorder of childhood and death in sanitized pleasure ghettos was an equally obvious stopping of the arrow; not an escape, merely an emptying of life. Age divisions serve few productive purposes in contemporary Western societies: they contribute to a system of unnecessarily rigid striations across the smoothness that life could be and, in doing so, they act as blocks to the communication of ideas and values; they set up false oppositions and they delay the young and debar the old from desire and power

SECTION III

Controlling Spaces: Creating Places

Everything we have said so far has assumed that there is no question that culture, or cultural geography, just *is* in its own right, or that there could be any such thing as nature which was not a cultural category. In this section we want to concentrate upon spatial outcomes as conscious constructions, emerging from acts of will. Here the desire and the power will be up front, not merely left implicit. But 'emerging from' is not exactly the same as 'caused by', and we want to consider the continuity between people and their environments which is best thought about using Deleuze and Guattari's notion of 'the body without organs' and its relationship with the 'desiring machine' or 'assemblage'.

For Deleuze and Guattari the body without organs (a phrase they borrowed from Artaud) is a product of schizophrenia, a body 'without parts which does everything through insufflation, inspiration, evaporation, fluidic transmission' (Deleuze, 1990, p.75). They clearly do not mean this literally but are drawing upon Guattari's clinical experience, which led to the knowledge that an inability to perceive oneself as fragmentary or discontinuous with one's environment and contacts is one of the manifestations of the condition to which we give the name schizophrenia (whereas other psychotics often see different parts of themselves as completely independent entities which are potentially invasive). Instead of slicing this off as a 'madness' which can be explained only by placing it beyond the bounds of the logical structure, their seminal work, *Anti-Oedipus* (1983), seeks to explain this continuity as another, and, for the times we are living in, serious way of perceiving the world. (In the same way that the willingness to chop it up in order to think about it is not normally regarded as lunacy.) Here they are clearly not simply talking about the connectedness of the parts of a functioning system (for in this case the metaphor of the body made up of organs would be appropriate, and is one which most of us are familiar with); instead we find the attempt to see the whole infused with all of its aspects. (If you have any knowledge about Hindu religion this idea is not difficult to think, where a deity like Durga is not segmented into or dominant over Parvati. Saraswati, Kali, Laksmi, but is all of them and numberless local mother goddesses at the same time as being herself, the feminine of Siva and divinity as abstraction. Even the idea of the Christian Holy Trinity is useful.)

But the body without organs produces and distributes intensities (not categories) though the experience of *profondeur*, which is conceived of as the way in which people *feel* space (as in vertigo), in what they call a *spatium* (Deleuze and Guattari, 1987). This spatium is a space in abstraction, without characteristics, which may then be *thought* as *extensio*, a measurable space with characteristics of depth, width and so on.

The meaning of space becomes located in the outcomes of the relationship between desiring machines, their bodies without organs and the 'nomadic subjects' which move across its surface. These nomadic subjects are the points of pure intensity which form what we regard as particular categories of the sort we were looking at in the last section, but they are never fixed, always in a state of becoming.

We admit that none of this rhizomic thinking (see Box X) is easy to think in such abbreviated form, but in this context it is not possible to expound further. We are certain that this section will be perfectly comprehensible without any profound knowledge of Deleuze and Guattari, but we would urge that they are worthy of attention, not only because they have directly informed the practice of major designers and architects such as Rem Koolhaus and spatial theorists like Paul Virilio, but also because they have been major influences in prioritizing desired production over mere profit accumulation in explaining power. It was Foucault (1977, p.165) who said, 'someday this century will be known as Deleuzian, and we are inclined to agree with him.

Obviously all spaces and constructions within them (physical and conceptual) are the outcome of these configurations of desire and power, but this is particularly apparent in those situations where the intensity is greatest, where an obvious statement of control is being made. For this reason we have focused on three topics. First, we look at spaces of confinement, order and surveillance Chapter 11, which draws heavily on Foucault's ideas about power and knowledge in disciplinary spaces like

BOX X Rhizome

Deleuze and Guattari (1987, p.7) argue that 'a rhizome ceaselessly establishes connections between semiotic chains, organizations of power, and circumstances relevant to the arts, sciences and social struggles'. Rhizomic thinking can be characterized as root-like, musical, like wandering around Amsterdam's canals, and contrasts with the overstructured, dualistic, aborescent or tree-like thinking with its neat definitions and oppositions that has characterized (white, male) Western thought for so long. The rhizome stops only when it confronts an obstacle or when it is ruptured. (Rhizomes are plants like orchids or irises.)

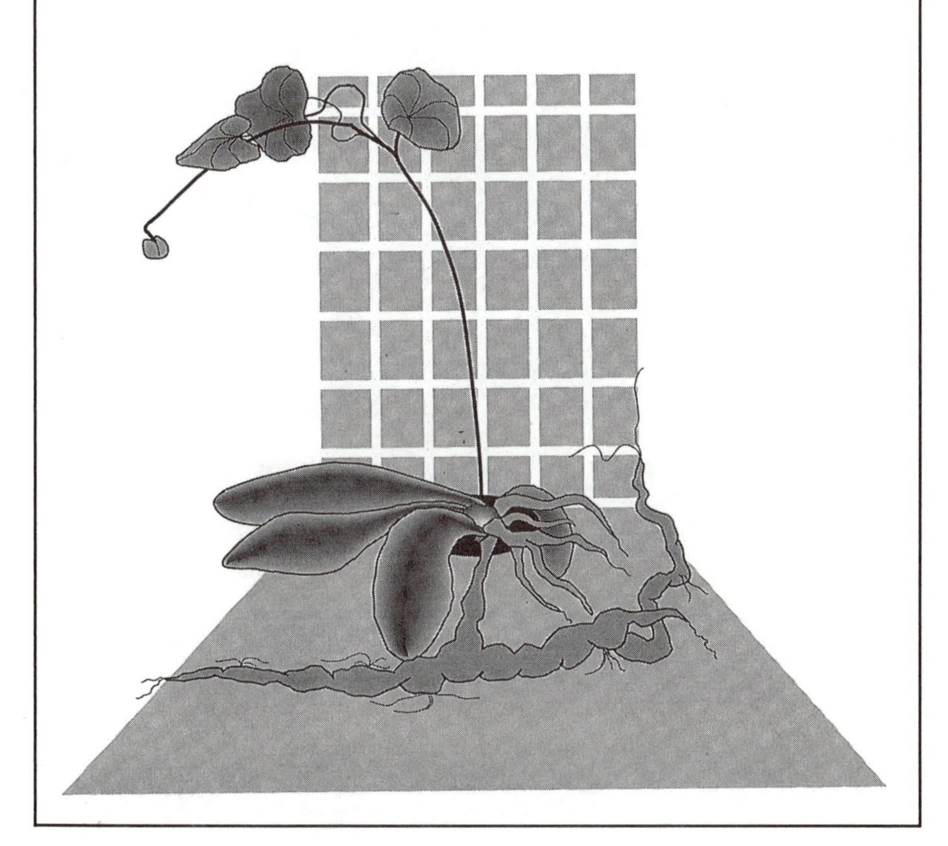

schools, prisons and asylums. Secondly, in Chapter 12 we move on to the commodification of space and the way in which places are constructed through a concentration of market forces. And finally, in Chapter 13 we look at the spectacular uses of space in the construction of monuments to abstractions of masculine power and desire, which then become instruments of control and mobilization through the sensation of awe. However, spectacles are also magnificent examples of pure intensity and sites of concentration of energy. They can contain within them transgressive and subversive elements which may serve as lines of flight.

11

Taking charge in modernity

In Chapter 9 we looked at the stigmatization of those in poverty, of those individuals segmented into an underclass with a range of pejorative meanings, most notably as health risks, and we termed this a power/knowledge relationship. The range of narrative discourses around the idea of the underclass set up a series of behavioural judgements and stereotypes. The most prominent writer to articulate the relationships between power and knowledge, discourses and practices was the French philosopher Foucault. His analyses examined the historical and spatial underpinnings of the stigma that we attach to those we term 'criminals', 'the insane', and 'the poor', and by reflection also what we mean by 'normality' in society. He argued that our notion of what it means to be an individual is a fairly recent invention – largely the product of eighteenth- and nineteenth-century thought. At the risk of oversimplifying, his overall argument was that during the late eighteenth and nineteenth centuries we shifted away from a society of control in which a sovereign or state governed territory or space *autocratically*, into an era when power relations began to take various *disciplinary* forms. Sovereign or state power focused directly on an individual's body as a form of expression. In the first few pages of *Discipline and Punish* (1977), Foucault describes in detail the public torture and execution of a murderer in France:

> 'Finally, he was quartered,' recounts the Gazette d'Amsterdam of 1 April 1757. This last operation was very long, because the horses used were not accustomed to drawing; consequently, instead of four, six were needed: and when this did not suffice, they were forced, in order to cut off the wretches thighs, to sever the sinews and hack at the joints. (p.3)

The spectacular public execution, and its description in various newspapers is designed to put a sense of fear into the hearts of the population, in short to control them through bodily terror.

Disciplinary forms of power which emerged after the developments in philosophy and theology resulted in the idea that a more subtle focus on the minds of individuals would lead to successful management of the population. In *Discipline and Punish*, Foucault goes on to describe the rules of a prison for young offenders in 1837: 'The prisoners' day will begin at six in the morning in winter and at five in summer. They will work for nine hours a day throughout the year. Two hours a day will be devoted to instruction. Work and the day will end at nine o'clock in winter and at eight in summer' (p.6). Here we find minute and inflexible attention to time being used as the framework of personal control, these timed activities would then have been confined to their own specific spaces and movement between them would have been highly supervised. Instead of attributing this change to a gradual humane, progressive force, Foucault argues that the emergence of disciplinary forms of power sought to spatially exclude and confine deviants from normality within specific institutions in order to produce a certain dominant idea of what normality was within the rest of the population. Whilst excluded, inmates, deemed to be those with insufficient regard for the structures of society, would find structure heightened almost to the point of parody.

Foucault's first book, *Madness and Civilisation* (1961), looked at the stigmatization and subsequent incarceration of those judged to be mad in the eighteenth and nineteenth centuries. Later he published *The Birth of the Clinic* (1973), an examination of the rise of medical discourses and their manipulation of ideas of health and normality in relation to the human body, followed by the previously mentioned *Discipline and Punish*, a study of the birth of the prison, as well as a number of other books and interviews.

These books are by no means straightforward, however. Foucault is a writer who never quite clarifies what he is saying; he deliberately mixes up his ideas and facts and identifies all the deviations and discontinuities in his historical analysis. He does this not out of arrogance, but because he wants to allow events to express themselves in what have been termed *spaces of dispersion* (Philo, 1991). He doesn't want to impose a false order on the events that he is studying, but allows the events he is analysing to speak for themselves, recognizing that such 'speech' is capable of many interpretations. Foucault's aim is not to set up another intellectual base-camp, but to explode the rationality of a number of historical events often taken for granted. Unlike Marxists, who argue for a single *metanarrative*, he argues that there is no single truth or essence to be found, only a series of truths: 'nothing is fundamental; that is what is interesting in the analysis of society' (Foucault, 1984, p.247). Nothing is fundamental in itself, but what is interesting is always the interconnection of events, processes and spatial strategies for an analysis of the operation of power within society.

Moreover, instead of just looking at who has power – the upper classes, the gentry or the politicians – he argues that we should be looking at how power is realized and applied; how we have been made into subject individuals by the operation of power through space. The idea of disciplinary power is conceived as positive, shaping and coercive, and differs from arguments based around ideas of domination and repression. Power produces constantly shifting rituals of truth, and is exercised rather than possessed (Matless, 1992). For example, with regard to sexuality Foucault argues that 'bourgeois culture does not repress sexuality, but through the spread of discourses on sex, including psychoanalysis, forms of sexual practice are created' (Poster, 1984, p.34). It is only after practices have been created and individualized that forces of operation can come into being. Thus he argues that the emergence of disciplinary power is a precondition for the success of processes of domination such as capitalism and patriarchy: 'Without the availability of techniques for subjecting individuals to discipline, including the spatial arrangements necessary and appropriate to the task, the new demands of capitalism would have been stymied' (Rabinow, 1984. p.18). In short, disciplinary power helps to make us into 'normal' individuals, relatively docile subjects performing our gendered roles in capitalist society: 'Discipline increases the forces of the body (in economic terms of utility) and diminishes these same forces (in political terms of obedience) ...' (Foucault, 1977, p.138).

Foucault outlines four ways in which power operates through specific discourses in order to make us 'docile' subjects; all of these discourses seem a matter of common sense, until one contemplates the relationships enshrined within them, recognizing the role of a stance of rational detachment in the exercise of power. We shall think of them predominantly in relation to a traditional school:

(1) There is a tendency for *hierarchical observation*. This has been commonly referred to as the power of the gaze or the surveillance of supervisory individuals within institutions. The prefect system of many British schools is a good example, or the dining hall with its raised platform for teachers to sit and survey the eating habits of their pupils, or even a teacher or lecturer walking around at the front of a classroom or lecture theatre. The web of educational surveillance is seemingly endless, yet it is easily transgressed through whispers and parody. In the film *Dead Poets' Society*, Mr Keating gets all the boys to jump onto the top of his desk, and 'take a good look around' in an attempt to subvert the formal spatial and observational power of the classroom through positional reversal. (But even in his subversion, he remains in charge, directing their transgression.) At the basis of hierarchical observation is the notion that the person who surveys does so as an incumbent of an office, not as an individual; the observing is framed as a duty, not a pleasure. Hierarchical observation extends into most jobs in factories and offices with lines of management and the layout of the workplace will reflect this, with those in supervisory roles accorded increasing privacy as they move up the hierarchy. Similarly, prison architecture is designed for maximum surveillance, through open stairs and galleries, peepholes, towers and security cameras.

(2) There is always a degree of *scientific classification*. Censuses and surveys permit the management of the population as well as its simple observation. The collection of case studies and statistics on such things as intelligence or health further leads to the scientific classification or segmentation of individuals into various formal categories such as 'the undeserving poor' or 'the mentally ill'. We have already mentioned Pratt's (1993) view that much of the scientific exploration in the early imperial era served as a process of power by classification. The gathering of data to establish the state of the whole population is a precondition for the successful management of individuals, even though the scientific process seems disinterested (Shumway, 1990). Again, in schools, this is shown in regular monitoring of physical, behavioural and intellectual 'progress', and a system of control by means of classes generated according to tested intelligence.

(3) All classification involves a number of *normalizing judgements*. These are vital for defining 'correct' forms of behaviour, generating norms. They may act through negative sanctions (punishments) or positive sanctions (rewards). So in schools, there are numerous rules for lateness, politeness and dress in order to standardize behaviour with boring repetitive penalties, such as lines, for departures from the norm as well as awards for conformity and good behaviour, spelling and handwriting (i.e., housepoints and commendations). Not all normalizing judgements, however, are based on formal rules, convention is also important, especially for establishing taste, attitude and ethos.

(4) *Examinations* are generally employed. These are highly ritualized ceremonies of power and means of establishing 'truth'. In schools, the spatial layout of the examination room with its ordered rows of desks, the strict ordering of time and the instilling of a fear of punishment if one even considers cheating are an example not to be overlooked (Merquior, 1989). Obviously within the context of the examination there is an asymmetrical relationship associated with knowledge. But there are also medical and psychological examinations, beauty contests, artistic competitions and the cross-examination of the criminal in court, all of which use authorized assessors. Examinations often then provide the results for further scientific classification.

All these forms of disciplinary power are clearly found in many state institutions and are often concealed behind legal frameworks. For example, in the case of the new prisons that emerged in the nineteenth century, legal judgements began to be concerned to establish not only that a criminal act occurred and who was responsible for it, but also why the act was committed so that the court could use this knowledge to determine the appropriate (rational, rather than vengeful) punishment. As Ogborn (1993, p.32) argues: 'A specific space is opened up in any legal framework for the intersection of law and discipline: the space of situated interpretation.' We thus began to have experts in court determining whether or not the accused was sane or if they were fit enough to stand trial. Of course, if he or she was found insane, though no crime against society was seen to be committed, the accused went into another disciplinary and custodial institution, namely the madhouse.

Foucault's ideas have principally been used by historical geographers in their analyses of the emergence and operation of a range of urban institutions in *modernity*, that is, the period of modernization and social change since the industrial and political revolutions of the eighteenth century. Driver (1989) outlined an historical geography of the workhouse system in England and Wales. Built to new architectural designs, these aimed to eliminate the moral threat of pauperism through training, supervision and segregation. There was a distinctive fashion for architectural design that allowed maximum surveillance, based upon Jeremy Bentham's Panopticon or Inspection-house design. The *possibility* of continual inspection was supposed to promote a fear of being caught in pursuit of those habits which were thought to exacerbate lethargy and perversions in the inmates. It was presumed that the fear of continual inspection and punishment would cause inmates to refrain from all misdemeanours, as Philo (1989, p.262) explains:

> The blueprint called for numerous single cells to be positioned on the radii of a circle, each of which was supposed to face inwards through iron gratings towards a central inspector's lodge from which it would be possible to see the actions of every inhabitant of every cell. In part, this visibility would be maximised through the spatial separation of inmates. Bentham also believed that this separation would inhibit the 'contagion' of bad thoughts, bad behaviour, and disease.

But, disciplinary space is not just about a series of panopticons. In a sense, it would be too easy to project an uninterrupted system of disciplinary control across a variety of institutions, from hospitals, prisons and workhouses to cemeteries and museums. But, in Britain at least, central state policies were interpreted at the local level before implementation. The aim of the state was to manage local differences through a new language of urban social policy, but there was, and still is, no such thing as a typical institution and some institutions provided moral training for individuals without recourse to such graphic spatial confinement. For example, Driver (1990) has described how the Mettray agricultural colony in France, provided a quite different model for the treatment of paupers. It was the direct opposite of a prison regime and sought to control individuals through social regeneration and a 'family system' of moral retraining which took the form of 'healthy interaction with moral landscapes and fresh air', but included a degree of surveillance.

The emergence of institutions was linked to new aesthetic discourses as well as moral ones. So the 'moral' landscape of the cemetery, whilst clearly a product of the concerns of the public health movement in

Britain, sought to remove death and pollution from the local parish churchyard and into the more aesthetically controlled space of the cemetery. In Portsmouth, for example, two new cemeteries were built outside the city to exclude the dead from the living on the basis of notions of hygiene and medical reasoning following the Public Health Act of 1848 (Hannam, 1992). The state of the country's churchyards had been widely blamed for outbreaks of cholera, but there 'were said to be moral miasmas corresponding to physical ones; moral filth was as much a concern as physical' (Driver, 1988, p.279). The cultural and scientific construction of the dead as a polluting category, led, via moral concern and Victorian notions of horror and abjection, to a purification policy for the city. This involved the expression of new power relationships across the city, shown through positive and healthy, as well as educative and aesthetic discourses. So the new burial grounds were to become garden Utopias, serene, sanitary, ordered and rationally planned. The designer of many of London's cemeteries, Loudon (1843, p.134) argued that 'Cemeteries are not only calculated to improve the morals and the taste, and by their botanical riches to cultivate the intellect, but they also serve as historical records.' In the nineteenth century, cemeteries, like many other institutions were moral, regulative and aesthetic statements in the urban environment (see Fig. 11.1).

Fig. 11.1 St Mary's Cemetery, Portsmouth

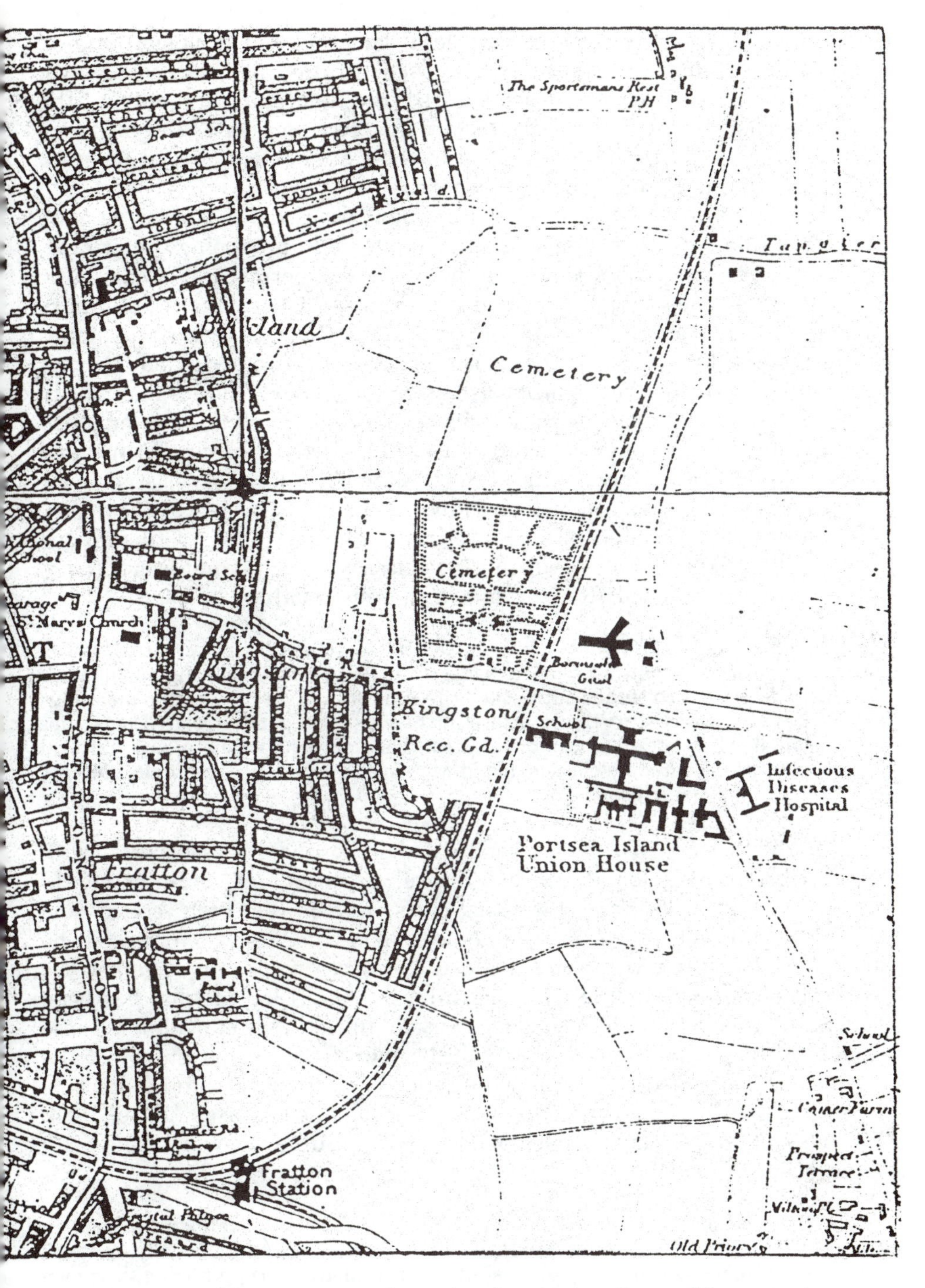

Fig. 11.2 Map of Portsmouth, 1890, showing the locations of the cemetery, workhouse (Union House), prison, school and hospital

As in many cities, ideas about the separation of polluting abnormality from clean morality resulted in a local geography of exclusion. Figure 11.2 shows the nineteenth-century concentration of the prison, fever hospital, workhouse and cemetery, with the lunatic asylum only a little to the east. All are aesthetically pleasing environments, and although there is no reflection of the stigma in their public faces, everyone knew the conditions behind the facade.

In describing these institutions though, we reach a point where we realize that though they are an effective force in producing conforming, docile subjects, people continue to transgress norms and break laws. Foucault argues that the apparent failure of institutional disciplinary power has, in fact, led to more diffuse and subtle disciplinary techniques that serve to justify further social exclusions. For example, former offenders have become subjected to disciplinary surveillance methods in their own homes, by 'tagging' and probation and social workers, and because of state and academic discourses causally linking crime and poverty. There is clearly a need to theorize some form of resistance, for resistances to norms can be found in virtually every historical document.

Mort points out in his book *Dangerous Sexualities* (1987, p.60), that one of the most subtle ways in which resistances were played out was through nineteenth-century discourses on health and sexuality:

> In the period of nascent industrial capitalism it was the systematic forms of knowledge of sanitary science, social medicine, evangelical religion and philanthropy which staked out a specific regime of sexuality. . . . Sexual meanings were constructed at the interface of the class and gender related polarities laid down by moral environmentalism. They marked out distinctive cultural identities for middle-class men and women and crystallized dominant images of the labouring poor.

In the context of fears of venereal disease and its association with prostitution and female sexual desire, a variety of mid-nineteenth-century texts described working-class women as both erotic and depraved. It was primarily middle-class men who 'used the image of the victimised wife and infected child to reinforce monogamy against the traditions of male infidelity' (Showalter, 1992, p.195), in what has been termed 'the rhetoric of medical terrorism' in the light of syphilis. These descriptions were buttressed by the women's reform movement of the time, which relied on an image of morally virtuous womanhood transgressed by the state legislation of the Contagious Diseases Acts of 1864, 1866 and 1869 (Poovey, 1990; Ogborn, 1993). But, as Poovey (1990, p.30) argues:

> Because prostitution *did* raise the issue of female sexuality in a form in which it could not be marginalized, these discussions paradoxically provided the discursive conditions of possibility for women's conceptualizing their own sexuality, and therefore the opening that would eventually enable women to

help change the way in which female sexual desire was represented and understood.

The notion of investment was crucial to this possibility, because it redefined power in such a way as to emphasize the agency of individuals in resisting dominant power strategies. As Deleuze and Guattari (1983) remind us, we often invest in the very systems that oppress us, not for disinterested reasons, but because of the possibility of a line of flight out of those systems.

The relationship of Foucault's corpus to feminism is itself a contested terrain, however, and many argue that the link between power and knowledge once again confines female discourses to the periphery (Balbus, 1987; Hartsock, 1990). But we begin to see that Foucault, in writing his histories, was, in fact, engaged politically in contemporary events (Eribon, 1989). He was no mere historian, but a philosopher engaged in writing histories of the present. As he put it at the end of *Discipline and Punish* (1977), these studies must serve as the foundations for studies of normalization in the present. He saw forms of disciplinary power embedded in our contemporary society, and geographers have recently begun to study these and the consequences for such issues as citizenship (Bell, 1993).

It has been argued that, whilst modernist disciplinary landscapes and institutions still exist, we are moving into a postmodern society and are in transition from the closed environments of discipline to a series of more controlled environments which intensify bodily desires. Former institutions of discipline have been renovated and sometimes perversely gentrified (for example, of Portsmouth's workhouse has now been converted into flats, perhaps untactfully reserved for single-parent families, see Fig. 11.3). A perpetual training of the body has replaced much of the traditional discipline of the school, and examinations have been increasingly replaced by continual assessment. Police forces have extended and refined their methods of detection, whilst remaining supremely institutional (see Box Y). If we work in a large company we can expect to be trained, retrained and monitored on courses throughout our career lives, Deleuze and Guattari (1980, p.112) argue that 'the means of exploitation, control and surveillance are becoming more and more subtle and diffused, in some way more molecular'. Deleuze (1991, p.4) further argues that whilst the proliferating new daycare centres, clinics, therapy, advice and guidance centres do express new freedoms, they also participate in mechanisms of control equal to the harshest of disciplinary confinements. Therapy and advice, a more subtle form of control through the power of confession, is widely available and in surrogate form it is mediated through advice columns in magazines and by daytime TV programmes such as *The Oprah Winfrey Show*. In a variety of ways we subject ourselves to examinations, all of which involve power relations that attempt to control and normalize our bodies, practices and habits; often we do this

Fig. 11.3 Portsmouth workhouse after renovation

in the name of self-improvement, surrendering ourselves into the hands of those with authority. It is the very fact that we frequently enjoy the personal care and attention involved that makes them so powerful in shaping our desires.

Sometimes it almost seems as if we are under perpetual camera surveillance, deceptively incarcerated in a series of striated spaces. But, as a case of child murder in Britain in 1993 showed, it is possible to record several stages of a crime on a series of video cameras without the crime being prevented or even recognized until it is too late. De Certeau (1984, p.111) has written of how even our movement on board trains involves feelings of incarceration:

> A travelling incarceration. Immobile inside the train, seeing immobile things slip by. What is happening? Nothing is moving inside or outside the train. The unchanging traveller is pigeonholed, numbered and regulated in the grid of the railway car, which is a perfect actualization of the rational utopia. Control and food move from pigeonhole to pigeonhole: "Tickets please ..." "Sandwiches? Beer? Coffee? ..." Only the restrooms offer an escape from the closed system. ... A bubble of panoptic and classifying power, a module of imprisonment that makes possible the production of an order, a closed and autonomous insularity – that is what can traverse space and make itself independent of local roots.

Our pleasure-driven modern Utopias have often quickly disintegrated into the harshest of dystopian confinements. Even such an overwhelm-

BOX Y The bridewell

The bridewell is in the centre of Newcastle. It has been in its present position and more or less in its 1977 lay-out since it was built in 1931. A hard, uncomfortable feeling pervades the place, but staff quickly acclimatize and come to accept their rough, spartan surroundings as normal. It is enclosed within a larger police and courts complex, set apart and deliberately built as a 'prisoner area' to exclude easy access for the outsider. Most prisoners arrive via a narrow arched entrance tunnel into an enclosed yard, from which a set of stone steps leads up to a heavy locked door with spy hole. A further locked door gives way to the charge room itself, which is the scene of furious activity on most days. Policemen, gaolers, detectives, prisoners, lawyers and others from the system mill around in what looks and sounds like chaos; great bunches of keys rattle while cell and grill gates clash as automatic locks snap into place. Yet there is always a pattern to the activity in this 20 ft square room.

The inspector sits on a high stool, a symbol of his power as the titular head, and works at an enormous brass edged desk over a dozen feet in length and 4'6'' in height. This is the 'charge desk' and it plays a central part in bridewell activity. It is scarred and marked from its fifty years of continuous use. Charge sheets, records of bail, and a record book with *particulars of every detainee cover its surfaces.* (Young, 1991, p.123) Young's book, *On the job* is an ethnographic account of police work from the inside. Young spent his whole working life as a policeman but also trained as an anthropologist.)

ingly fun place as Disneyland is a site of control, surveillance and normalization. Eco (1987, p.48) has called it a degenerate Utopia:

> Access to each attraction is regulated by a maze of metal railings which discourages any individual initiative. The number of visitors obviously sets the pace of the line; the officials of the dream, properly dressed in the uniforms suited to each specific attraction, not only admit the visitor to the threshold of the chosen sector, but, in successive phases, regulate his [sic] every move ('Now wait here please, go up now, sit down please, wait before standing up', always in a polite tone, impersonal, imperious, over the microphone). If the visitor pays the price, he can have not only 'the real thing' but the abundance of the reconstructed truth.

Eurodisney has run into financial problems precisely because of a cultural resistance to this degree of control in situations where one is supposed to be having fun. The abdication of autonomy in the name of order and safety is not rated as highly in a European context as it seems to be in America. The blandness, tidiness and routine seem sinister in a European setting with relatively fresh memories of Fascism, as do the snaking, martialed queues. One can enjoy it only reflexively, as a post-tourist, aping Eco's ethnography. Whilst we have argued in Chapter 4 that Utopias may be potentially revolutionary, all too often they repeat some of the worst features of contemporary life when they are territorialized. Somewhat paradoxically, it is the risk of dystopian elements in society that these palaces of fun conceal in their imaginary and regulated spaces; they exclude all the dangers of Western overconsumption in their drive for official pleasure.

Beck (1993) has linked this to the emergence of a *risk society*, but it was Douglas and Wildavsky (1983) who first brought our attention to the meaning of risk within societies. We can see how banks and building societies now trade information on individuals who are assessed for their credit worthiness, or rather their credit and health risks. Security cameras have become an omnipresent feature of many middle-class homes along with various burglar alarms and multiple locks for both insurance and deterrence reasons, again, in order to minimize crime but also because of an increased awareness of risk. Insurance itself assumes a gambling on the basis of perceived risk and introduces the notion of the very real fear of the uninsurable risk, which is more like a threatened danger. Whilst there may be a *need* for such devices, the *desire* for greater safety can be viewed as an extension and routinization of the power of normalization linked to the illusion that risk can be eliminated by a combination of surveillance and management. The security cameras we have already mentioned are there to facilitate the exclusion of those who do not look or act the part, loiterers in shopping malls, thieves in offices, but also to make sure the area symbolizes safety, that it is clean and pure, physically and morally. Through a panoply of mirrors they reflect good behaviour, instill normalized habits in individual shoppers, workers and children alike and minimize financial risks.

We may accept the notion of risk in society but risk is a construction based upon knowledge. Through our language we cope with the perception of the increased real and imaginary risks of living in a postindustrial society. Methods of surveillance and the use of a variety of speech-acts both seek to emphasize the triumph of safety over risk, normality over abnormality through the construction of binary opposites that can deny the existence of the real or imaginary threat, Zonabend's (1993, p.2) recent study of the 'nuclear peninsular' in northern France has explored these notions. She argues that:

> any questions about danger incurred or risks run will be rejected, denied or parried in some way. People who interview populations living in the vicinity of nuclear power stations are well aware of this phenomenon; every poll ever published shows that, the nearer people live to a nuclear power station, the

> more they will swear by its reliability. Similarly those who observe workers in high-risk industries are familiar with the way in which they refuse to acknowledge the dangers of their job to the point where it is hard to get them to admit to taking essential safety precautions.

Yet at the same time, 'This game of life and death is one that the men of the nuclear industry rehearse daily in their words and in their actions' (Zonabend, 1993, p.123). Just as at Disneyland, there is a sense of seriousness in the play, as risks are both courted and calculated. Many men who have been exposed to radiation use a vocabulary that emphasizes either their heterosexual virility, or their worthiness in battle, in order to cope with rumours that they could no longer be 'real' men. Women, in contrast, avoid naming radioactivity, admit to being scared of it and exclude it from their vocabularies. Men often draw on an idea of tradition; they compare themselves to coal miners in a claim for heroism, risking death for a wage. Yet, unlike coal mines, the risk of living near to a nuclear site should not be associated solely with the men and women who work there, but also with their families and neighbours. Feelings of punishment for involvement are widespread, more diffuse. In contrast the more specific forms that Foucault outlined, Zonabend notes:

> Their punishment may take many forms. It may affect them more directly, or it may strike at their families. It may manifest itself immediately or at a later date, long after they have ceased to work at the plant. For some the punishment may never come. For others it lies in the future. For one or two it is there already, taking the form of a cancer, an illness that is often seen in terms of punishment undermining a person's physical integrity. (p.118)

She concludes that, 'all their various strategems are so many tricks and tactics for taming the dangerous, unknown world in which these men and women not only have to live but need to go on living' (p.125).

Spaces of both high risk and high fun, it would seem, are normalized through both the language of play and the fear and practice of punishment. We saw how in the nineteenth century a variety of disciplinary institutions emerged which sought to normalize the practices of sections of the population through such categories as insane, criminal and delinquent. By excluding those categories from 'normal' society it was thought the risks of contamination associated with these groups would be minimized. But this also had the effect of controlling the practices of the rest of the population through both feelings of guilt and a fear of punishment. Today, we can see larger and larger areas of Western cities and countryside being transformed into either manageable risk zones, or managed fun zones, through both technological and political developments. Insurance companies regularly designate certain cities, and even areas of cities, as well as individuals, as high-risk, and therefore high premium; commodifying abnormalities in both space and people's bodies in an ever increasing web of control.

12

Commodified spaces

Some readers probably wondered at the title of this chapter; if space is an abstraction, how can one commodify it? Surely it was just an intellectual slackness that caused us to write 'space' not 'place'? Surely this chapter should have been found in the first section of the book, looking at places which can be bought and sold, not the section which focuses on control? We are hoping to show that when places are being offered for sale, it is space that is being commodified and we want simultaneously to demonstrate that commodification is not a simple matter of putting prices on things, it is also an important ingredient of power in that it is selective, exclusionary and regulative. Commodification is not just about the market, it is also about the process of accumulation of benefit through its alienation from the realm of the unpriceable. We find ourselves once again returning to the theme of the two economies of scarcity and abundance, want and desire, and a congruence between economic alienation and existential alienation. This form of accumulation is one of the major props of centralized power and we argue that, as more aspects of life become commodified, more and more regulation takes place and less remains free. The word 'free' can imply either personal liberty or that which is available outside the market, without price, and there is an obvious, but often duplicitous, elision between the two. In geographical terms, this can be experienced in the exclusion of non-payers from formerly 'free' spaces, in different ways of valuing (in both senses) space, place and environment and, quite simply, in changed ways of thinking one's being in the world.

It was Marx (1974) who alerted us to what he called 'commodity fetishism' in the first volume of *Capital*. We can see this as the way in which a *mere thing* takes on a spectrum of values as soon as it is commodified. A commodity being something (including services and attributes) which is buyable; it is subject to market forces, with all the morality and politics which adhere to these. People encounter commodities in very different ways from those in which they encounter non-commodified things but there can be a crossing over and blurring of the divide between them in representational and metaphorical practice. We can see this particularly clearly in the gift economy, where things that are taken from the market economy enter into the realm of sentiment and then achieve a new value; this is often perceived as superior to the mere market value, but a gift from someone who is despised will become a tainted possession and may be lightly disposed of. It is also apparent in the case of inherited goods, which can become imbued with the sentiments associated with the bequeather.

The major difference between a commodity and a non-commodified thing is that its value becomes overt, not implicit. We may think of commodities as having an *exchange value* in addition to their *use value*. *Labour value* becomes apparent only once one introduces the concept of exchange value, as labour seeks to apply itself to those situations where the exchange value of the commodity is greatest. However, labour itself becomes commodified in this process and not all people's hours are then valued in the same way. In simple terms, this means that a price can be established for any good, though at any one time the price may not be known because its commodity role is lying dormant (it is not up for sale, though it is saleable). It should be obvious that commodification is a cultural and political matter as it indicates a shift of 'free' things into the economic domain and creates scarcities out of abundance. This has implications concerning entitlement and appropriation when we find that things which were formerly corporately enjoyed or freely given between individuals need to be bought and thus become difficult to obtain. Accounts of absolute privation, as in Poewe's (1988) experience of Germany just after the war, dwell upon the commodification of everything – paying to be placed on a list for accommodation or transport, selling sexual services for basic goods – blurring one's notions of family responsibility and corruption. Kristeva's (1982) notion of the abject turns upon the idea of the self or the thing which is experienced as worthless, in an obvious (and deliberate) confusion of spheres of value. Whenever something first makes the transition from *thing* to *commodity* there is often confusion, incredulity or just plain resentment at the point at which mutually exclusive value systems intersect.

For example, when Lewanika, king of the Barotse, 'sold' the mineral rights of much of what is now Zambia to the British South Africa Company for £2000 a year he did not 'own' the land in the sense that a capitalist landowner does, nor did he have any notion of the possibility of the alienation of rights over that land, belonging as he did to a society in which people could hold only the rights of usufruct over communal land (Caplan, 1970). (Usufruct implies the right to the product of land, but not the right to dispose of the land. A person who holds rights of usufruct can prohibit other people from using the land only so long as it is being used. This means that the land is not owned, but it is giving benefit. The French term *jouissance*, which we have already used several times, has usufruct as one of its meanings.) The Barotse story is one that can be told thousands of times over in colonial history. Seemingly a tale of canny operators driving a shrewd bargain and persuading simple people to part with valuable resources for paltry sums, it is actually far worse than this in moral terms, for what was actually happening was that entire local systems of value were being overturned by a market system which they had no tools to conceive of. People's relationship to the land they used, and to each other, had to change once it was alienable, for processes of exclusion set in immediately land can be bought and sold. Non-market values then retreat into a subordinate realm of 'tradition', guarded predominantly by sentiment, but eternally under siege from market forces once commodification becomes part of experience. Only when tradition is commodified as 'heritage' does it become revalued, but this is full of obvious contradictions. We may think of this as the construction of blocks and barriers to the flow of benefit; these are then reinforced by legal systems and structures of social differentiation in a move from the smooth to the striated.

In Britain we are most conscious of commodification in the name of 'privatization', associated with the New Right; this is the process whereby things shift from the public realm (that is, from corporate ownership, via the state) to the private realm (ownership by companies, ostensibly made up of individuals) where profit will be taken. We can also see market principles at work in the setting up of a multitude of 'authorities', 'trusts' and 'boards', quasi-public bodies which act as cost-centres contractually supplying services to the local state, competing with each other to minimize cost but often doing so at the expense of the service rendered. The effect of this is to make people aware of the economic value of things which were formerly thought of as a civil right and to normalize the exclusion of sections of the community from access to them. Indeed, notions of rights and community start to look rather threadbare once one crosses over into the realm of purchase, for there is a very real difference between equality of opportunity and the right to be treated equally by market forces. This then brings into question the issue of citizenship of those people who are not merely poor, but come to be categorized as an underclass which is seen as undeserving because it cannot pay. The changing morality implied by a move to an enterprise culture was examined in a collection edited by Heelas and Morris (1992, p.2) who saw it as 'designed to change how people think and what they value. Fundamentally it is to do with changing how all of us understand ourselves ... the culture is perhaps *the* vehicle in the attempt to establish a regime where the individual, rather than the state apparatus, can flourish'.

Creeping commodification has altered concepts of the person in those societies dominated by market exchanges. Magazines and advertisements relentlessly urge us to improve the value of ourselves and our possessions (particularly the owner-occupied house or flat) and the value which is emphasized is that of the market. Saunders (1986) showed us how those people who consume predominantly in the private sector of the economy (i.e., those who buy their own homes, pay school fees and medical charges, drive their own cars, have a large range of consumer durables and are not dependent upon welfare services) have very different attitudes and behaviours from those who consume predominantly in the public sector. Communitarian ideals are eroded and there is a resentment of taxation and public spending; life becomes an investment. They start to resent the provision of 'free' benefits to those whom they deem unworthy of them – in comes means testing, the issue of entitlement and the inevitable construction of 'outsider' groups which come to be seen as 'underclasses'. Gradually everything comes to have a money value – a seat in the park, a view, admission to a museum, a swim in the sea, a walk in the countryside, a burial place, to name just a few of the things which we have sometimes encountered free of charge and sometimes not.

However, privatization is not the only form of commodification; it takes place in the shift of behaviour in relation to goods and services which occurs between ordinary individuals. In Chapter 7 we talked about the way in which generosity was not predicated upon having a great deal of wealth, but in having a concept of abundance; we might venture that abundance and generosity are perhaps best thought of as valuing goods less than relationships. (Formalist economic anthropology which sought to explain gift-exchange as a mere economic investment in human relationships, designed to increase the return of goods, missed the point about differently constituted values. Many years ago Pam published a piece on Christmas gifts [Shurmer, 1972] and upset all her friends and relatives with her cynicism. They knew much more about the separation of the two economies than was recognized by conventional anthropological theory at that time.) With commodification, things, including time and services, acquire a monetary value which takes precedence over human relationships (this can apply at any scale from the intimate to the global). Simmel alerted us in *Metropolis and Mental Life* (1971) to the way in which money and the clock, in tandem, quantified human relationships in urban settings and, in doing so, could drive away sentimental attachment like friendship. His major work, *The Philosophy of Money* (1990) elaborated this into a whole way of looking at life in modern times and, though it is one of the great classics of sociology, it has been rather neglected by geographers, who have been more exposed to his disciple, Louis Wirth. Simmel perhaps overstated the case, in his assumption that urban life was a cause of mental disorder, and it is easy to be taken in by the altruism which always seems to characterize premarket exchanges, but there is a very real sense in which a growing commodification does infect all value systems. Most of us are hurt when a guest makes an estimation of the current market value of our home or comments on how much we must have spent on hospitality. Somehow we feel that our generosity is tarnished if people value it with the money which is designed to facilitate the taking of profit, rather than with the sentiment which underlies direct human encounters. A guest who talks in market terms about what we provide almost becomes a 'punter', who seems to be able to decide whether a good bargain is being struck, who could apparently pay the bill and, without any feeling of responsibility or guilt, discontinue the relationship. Friends have different obligations from customers and these are not contractually agreed. With increasing commodification more and more often the thought will cross one's mind that one is not getting a sufficient profit out of one's personal interactions – after all time is money.

In political terms we can see the ethical separation of the market from the personal in the concepts of bribery and nepotism and the requirement that economic interest should be declared. Scandals over the illegitimate fusion of political or personal contact with market position, as in insider dealing, only have any meaning so long as one believes that the market should be subject to controls, that it should not be 'free' to operate entirely in its own interests. This belief becomes increasingly difficult to maintain with a move to an enterprise culture with a weak concept of society. We can see a change in the nature of commercial relations as the formerly personal element within them becomes commodified. At the most banal level this means that human contact with a waiter or air stewardess is shut off in formulated friendliness ('I'm Lucy, how may I help you . . . have a nice day'). In geographical terms, however, contact commodification is more significant in those situations where local interpersonal ties, of the sort which operated in the traditional City of London, give way to globalized unidimensional links which are electronically forged; here the time involved in a transaction can be precisely valued in the profit made, as well as the money spent on the communication. Emotion and non-market value are increasingly eliminated.

With the increasing importance of market values, ideas about society change; at all levels up from the individual to the global an *economizing* sets in. A significant aspect of this is that the individual is encouraged to perceive him or herself as a site of investment. Body maintenance and improvement, and personality development become important to one's career chances, but one's life begins to be seen as a career itself. The person comes to be a valuable construction, rather than a given; a body becomes something like a possession or a display of the worked-upon self. (On the subject of the commodification of the woman's body, Wolf's *The Beauty Myth* [1990] makes chilling reading, giving numerous examples of discrimination against women in employment on the basis of their looks, even in occupations where beauty is not a logical prerequisit.) Beyond the individual level, relationships begin to be subject to an implicit cost–benefit analysis and notions of friendship, neighbourhood and community show signs of strain. In Britain in the 1980s (the time of the most significant shift from corporatism to market orientation) there was a quite deliberate cultural move on the part of the Thatcher government to instil 'enterprise values' in the minds of the people through a variety of discourses. This heavily relied upon the notion of a share-owning democracy, but also included the introduction of programmes of enterprise education into the school and university curricula. More generally, there were aggressive moves to foster a new selfishness and create the belief that individuals should take responsibility for their own life chances.

Taking care of oneself had a tendency to mean abdicating communal responsibility, but moralizing this in terms of self-reliance through a belief in just returns on one's own investment. Charity, dependent upon donations to those considered worthy of support, was urged as superior to welfare entitlement. Charity itself became commodified in great mediatized events which mobilized local activity, focusing on individual effort and competitive achievement, with a mixture of exhibitionism and altruism involved in these. Recipient groups are beginning to mobilize in opposition to their new status, as outsider recipients of charity, rather than normal members of society with entitlements. Groups like the British Direct Action Network deliberately embarrass wealthy 'donors' by organizing disabled people to picket outside exclusive and expensive 'charity' events, such as dinners and film premieres, and fund-raising 'telethons', which serve to enfeeble those with disabilities. We may see this as a recognition of the process of abjection and a rejection of the classification which would render one abject and stigmatized.

The very idea of community undergoes a transformation under conditions of increasing commodification, where it can become a ritualized *communion* rather than a functioning interaction. The spectacularization of charity in mediatized participation events is a major example of this. Berking and Neckel (1993, pp.71–2) consider the phenomenon of the urban marathon as an event which simultaneously personalizes a city and ritualizes relations within it:

> the marathon is the collective event of those individual strivings that require the city as social space. People run for themselves and against themselves, struggle for success with and in front of others, a success which is considered in the modern marathon to be the mere overcoming of the course. In its concentration of all the respective individual goals into a competition in front of the admiring mass audience, the marathon symbolises the intra-urban competition of contemporary achievers.

Marathons are important mediatized public performances and contain within the ranks of their competitors those who are running under charitable sponsorship, often signifying their lack of competitive seriousness by wearing fancy dress. Such events can be contextualized within the much vaunted shift in Western countries from industrial to service-based economies.

This change implies an increasing commodification of intangibles, in a logical extension of the prior commodification of goods. But personal services are very different from goods in that they have a human element and reside in the realm of experience rather than acquisition. The mixture of sense of freedom combined with moral opprobrium which is commonly attached to prostitution, and the sites it colonizes, has for centuries illustrated the ethical difficulty attached to the sale of personal services, for it is commonly assumed that these should reside within the realm of the freely given. However, for those already able to satisfy their material needs and wants, wealth has become a matter of the progressive heightening of experience or sensation and it has simultaneously come to be led by desire, as lifestyle and consumption are conflated. The market is looked to in order to provide services which would previously have been performed at no cost by known individuals but which would have been within the context of personal obligation and debt; they would not have been called 'services' had they not entered the market sphere. Bereavement counsellors, colour consultants, even psychotherapists could not exist without a belief in the possibility of the commodification of therapies for the wounded or incomplete Self, as Rose (1990, p.142) demonstrates:

> The guidance of selves is no longer dependent upon the authority of religion or traditional morality: it has been allocated to 'experts of subjectivity' who transfigure existential questions about the purpose of life and the meaning of suffering into technical questions about the most effective ways of managing malfunction and improving 'quality of life'.

The advantage and the disadvantage of this shift is that one does not have to put so much effort into the maintaining of webs of personal relationships – one becomes independent but isolated in one's 'freedom'. As the ties of community loosen, those who cannot afford counselling and therapy may well find themselves outside care. These new services are often provided from domestic bases or else are mobile, and visit the consumer in his/her home; many can be provided by telephone, audio-cassette or video, meaning that no recognizable space of personal service emerges.

Baudrillard, as early as 1968, drew our attention to the fact that consumption is a symbolic activity and, working from his Marxist beginnings, he gradually came to realize, through the 1970s, that a culture orienting itself towards sensation is not usefully analysed according to the principles of maximization of satisfactions grounded in the material realm (see Box Z). Goods become valued less for their material qualities than for the messages they can communicate, but even this is too mechanical, for messages become diffuse and symbols come, in the words of Wagner (1986), to 'stand for themselves' in the economy of desire. We pass then into the realm of the *hyperreal*, where 'representations' have a life of their own and refer back to nothing other than their own power to

BOX Z Baudrillard on consumption

We must clearly state that consumption is an active mode of relations (not only to objects, but to the collectivity and to the world), a systematic mode of activity and a global response on which our whole cultural system is founded.

We must clearly state that material goods are not the objects of consumption: they are merely the objects of need and satisfaction. ... Neither the quantity of the goods, nor the satisfaction of needs is sufficient to define the concept of consumption: they are merely its preconditions.

Consumption is neither a material practice, nor a phenomenology of 'affluence'. It is not defined by the food we eat, the clothes we wear, the car we drive, nor by the visual and oral substance of images and messages, but in the organization of all this as signifying substance. Consumption is *the virtual totality of all objects and messages presently constituted in a more or less coherent discourse.* Consumption, in so far as it is meaningful, is *a systematic act of the manipulation of signs.* (Baudrillard, 1988, pp.21–2)

represent. The media is vital in the constitution of hyperreality, but the hyperreal is also to be experienced materially in those sites of commodification which have appropriated a system of communicative value. Here we are concerned with the *simulacrum*, the fake which, because it does not imitate anything, is pure sensation and meaning.

From this point of view we can see a further distancing of things from their use value as the experience of consuming becomes more valued than the consumption. So we find that, for some people, shopping becomes valued whilst the goods bought are no more than a vehicle for the experience. For any human geographer this is a vital message to take on board as the culturally constituted (but constantly shifting) value system acquires more explanatory power than any other.

The most obvious way to think about this is with reference to the contemporary shopping mall which, though it has great similarity to the luxurious and exotic grand department stores of the end of the last century, has practically nothing in common with the super- and hypermarkets of modernist mass consumption. Malls are seductive not only for shoppers and flaneurs but also for academic geographers and sociologists, who have written at such length on the subject that it is becoming difficult to think of new phrases to describe them (Shields, 1991a). Rather than being simple points of supply, malls aim at being temples to consumption; what is offered up is the shopping experience, which is paid for through the purchase of goods. Once one makes this leap in one's thinking, there is nothing surprising about the environment of excess and the sensory overload of a successful mall; there is no reason why there should not be a waterfall and a grotto, allusions to aristocratic living and the oriental market; why there shouldn't be lush vegetation and a hint of a Roman palace; why we should wonder at the way in which shopping, play, eating out and being entertained flow smoothly into one another, evoking the traditional carnival or fair, rather than the modern high street.

We should not be fooled into thinking that the mall is functionally the same as the carnival, it is merely carnivalesque. Where the carnival is a celebration of community, the mall is a celebration of consumption – and the obvious route through which such a celebration would take place is that of its commodification. Unlike the carnival, which has transgressive and oppositional aspects (Stallybrass and White, 1986), the mall is exclusionary. Though it is internally relatively unstriated, in that the various individual shops have deliberately blurred their boundaries with the areas of circulation, there is a very real discontinuity between the mall and its surrounding environment. One enters into another world, a world of simulacra, which shows up the drab ordinariness of the mundane 'reality' on the outside. Part of the 'niceness', cleanliness, orderliness and luxury of the mall depends upon very high levels of surveillance and security and rigidly enforced codes of behaviour. A mall has failed, and will consequently lose its 'quality' retailers, when it cannot exclude the derelicts (or even the unfashionable) and when the war against litter is lost. It relinquishes its role as a purveyor of consumption and becomes a mere centre for the distribution of goods. The *Forum les Halles* is a well-known illustration of this, intended to be an internationally attractive centre for very high-quality retailing, it has drifted progressively down-market. In spite of its much publicized design and its famous site, a lack of natural light and an overidentification with its *Metro* station, meant that it was never quite able to establish itself as an élite or exclusive environment. It is now becoming notorious as being attractive to semi-professional shoplifters because of the ease of the getaway offered by the many escalators leading to the integral rail interchange.

It may be sacriligious to refer to *Madame Bovary* (Flaubert) as the first 'sex and shopping' novel, but we can see a progressive association between libidinous and commercial expenditure in the presentation of goods. This association has been picked up in literature and film as an erotic device, but in marketing it is employed to enhance the commodity. The recognition of the sensuousness of luxury shopping is clearly marked in the work of Benjamin (1970, 1979) and echoed by the neo-dandy intellectual *flaneurs*. Goods in the *grands magasins* of the late nineteenth century and in the malls of the late twentieth century are displayed to seduce the eye, to give an illusion of sumptuousness and luxury which they will lose when they leave their commercial context to become household utensils rather than commodities.

We should not be fooled into thinking that, because displays are sumptuous, the goods embedded in them are necessarily special, either in the *fin de siecle* Parisian stores or the contemporary malls. In his paper, 'Learning to consume' (1993), Laermans refers to the presentation of goods as a *mise-en-scène* (a term usually used in reference to film, implying that there is a deliberation in the setting). He maintains that women of the emerging *petite bourgeoisie* of the nineteenth century were educated in 'good' taste through stores like *Printemps* which cultivated a spectacularization of the commodity through the construction of Orientalism (as a metaphor for opulence) and Parisianism (as a metaphor for *chic*). The shoppers were not particularly wealthy; the aim was to cultivate in them a taste which could be met through the accumulation of mass-produced, or mass-imported, items. Laermans maintains that, 'Commodity aesthetics had to produce [the] very meanings that the commodities themselves usually did not express' (p.93). ('Commodity aesthetics' is a term we owe to Haug [1986, 1987], who was interested in the way in which value could be constructed, via artificially manipulated notions of taste, to construct meaning around intrinsically banal objects, like denim jeans.)

Much of the service provision of the postmodern era has been in terms of the appropriate aesthetics and has understood the hyperreal desires of its ideal clientele. Services are placed in a context of communication which is experienced as a fantasy in which the customer colludes. So swimming pools become tropical lagoons, not just tiled swimming areas with changing rooms; banks look like the foyers of traditional hotels; hotels look like Mexican villages or stately homes; offices pretend to be domestic apartments. Where modernist design attempted to deepen the legitimate meaning of an activity by emphasizing its function in its form, much postmodernist design seeks to proliferate meanings and to slide one realm of experience into another. (Where purists dismiss this as superficiality and stylistic crassness, they are missing the significance of the shift in economic value which accompanies this commodification of service.) The practice of scheduling corporate hospitality and important meetings in novel settings such as yachts, museums and ski resorts is another manifestation of the same process of divorcing form from function in a bid for new experience and a blurring of business and pleasure. This then opens fresh opportunities for commodification of private or exclusive environments, as museums open galleries as dining rooms and castles set aside conference facilities.

The more advanced stages (if we can use a term so heavily and misleadingly endowed with evolutionist connotations) of commodification aim at *ways of being and becoming*, not just levels of accumulation. The term 'lifestyle' captures this and lifestyles can themselves be bought and sold, not only their settings, costumes and props, but also the scripts and the 'selves', governed and therapized, that go with them. Shields' edited work, *Lifestyle Shopping* (1991a), which looks at contemporary commercialism from designer fashion in Tokyo to 'alternative' marketing at Stonehenge, is a compelling collection which captures the movement between modes of shopping as an *attribute* of the lifestyle of categories of customers, through the idea of shopping *for* a lifestyle, to that of shopping *as* a lifestyle. As this happens the categories, such as social class, which have long provided the certainties of social geography start to collapse and 'lifestyle' inscribes itself on urban and rural landscapes. Most obviously this is to be observed in terms of housing, as the various elements of what is problematically called the 'new service *class*' pull together lifestyles which are both congenial and communicative, assembling cultural capital and financial capital simultaneously. Such people are an important part of Western society and, in addition, their choices have implications for the value and values of others.

Housing has long been commodified, and it is not only in contemporary society that we have had an awareness of an additional value adhering to some districts and dwellings which has more to do with communicative than inherent use value. Gold and Gold's (1990) account of the marketing of 'Metroland' (the new housing to the north-east of London) between the wars shows how mass speculative housing was promoted as semi-rural traditional, amidst a whelter of symbols of health, prosperity, stability and substance. But it is only relatively recently that lifestyle attributes have been self-consciously attached to residential accommodation and marketed in blatant fashion. Mills (1988, 1993) shows how developers in Vancouver transformed Fairview Slopes, a run-down inner-city area of rooming-houses occupied by people with 'countercultural lifestyles'. They did this by playing upon the image of the young urban sophisticate who dreads being associated with the family-dominated mores of suburbia; they offered 'an *extraordinary*' lifestyle', 'an *exciting new* lifestyle', '*designer* living' but also emphasized the geographical contrast with the suburbs – '*urban* living', '*downtown* style living' – and they even recognized the sociological awareness of their potential customers by advertising 'an *upwardly mobile* lifestyle'. What they were offering in material terms was compact, easy-to-maintain apartments, close to the centre, but these rational considerations were far less promoted than the attributes which adhered to them – the freedom, the youth, the difference. It was out of these that the *investment* side of both development and purchase would be constructed. The place needed to take off in lifestyle terms for there to be real profits, and larger developers with established reputations for élite housing held back from the district until they were sure that the cultural as well as physical transformation had been made. We then have to recognize the existence of both gentrifiable neighbourhoods and gentrifiable people; it is not just a matter of a certain type of person moving in, but a conviction that a bid for a particular identity could be made via the move. Much of the problem with the London Docklands development was precisely that it could not obliterate the class antagonisms and long-stigmatized identity of the area (Bird, 1993; Dunn and Leeson, 1993).

We may see the idea of lifestyle as an aspect of Bourdieu's (1971, 1984) concept of *habitus* (see Box AA) as a means of distinguishing concentrations of behaviour, taste and consumption which were based upon more than simple class position or wealth. He was concerned with the way in which cultural (See Box BB) capital and economic capital entwine around and separate from each other to generate different cultural configurations in contemporary society. Bourdieu's analysis rests upon a highly specific ethnography of France in the early 1960s and can seem quaintly dated, preceding as it does the massive shift in the structure of the global economy; it does, however, offer pointers to the way in which consumption categories isolate themselves in opposition to one another in a heterogeneous society and he threw into sharp focus the importance of taste, rather than simple wealth, in establishing boundaries, policed by contempt and disapproval. For example, the people who were referred to in the 1980s as *Yuppies* (a word playing on associations of age [and family cycle], the urban and upward social mobility), were the creators of, but also created out of, what came to be popularly known as a *lifestyle* movement, based around the idea that one's style of consumption could both indicate and influence one's position in a new social and cultural order. Aware of lifestyle considerations, the geographical choices made by people begin to make more sense than if one confines oneself to mere utility and convenience as a basis of interpretation. We can understand the choices of those who are in a position to choose accommodation by taking account of the perception that lifestyle categories have of each other. Each location commodifies that difference and sells not only neighbours with a convergent habitus but, perhaps even more importantly, an absence of neighbours with a conflicting one.

Cooke (1990) and Relph (1987) have shown how 'pioneering' residents of inner-city developments have created boundaries around their colonies. Sometimes these are walls with security guards on gates, other times they are more or less a symbolic inward turning. Such settlements become reminiscent of the cantonments of the British Empire as a host of new service provisions flow into the area to cater to the tastes of the colonists in a move in the direction of institutional completeness.

Gentrification is not just an urban matter. In many European countries, but particularly in Britain, the blend of natural living, aristocratic superiority and peasant authenticity which is supposed to characterize the rural areas has become a valued commodity. The countryside has

BOX AA Habitus

First outlined in his *Outline of a Theory of Practice* (1971), Bourdieu's concept of *habitus* is most fully explored in his book *Distinction* (1984). Of particular interest to geographers is Chapter 3, 'The habitus and the space of life-styles'. The concept has not been without its critics (see Honneth, 1986) who have seen it as a both tautological and reductionist model.

For Bourdieu habitus is 'the internalised form of class condition and the conditioning it entails'. It is thus dependent upon the construction of 'the objective class, the set of agents who are placed in homogeneous conditions of existence imposing homogeneous conditionings and producing homogeneous systems of dispositions capable of generating similar practices; and who possess a set of common properties' (p.101). At its most banal, this means that habitus is the way that people who think and behave the same way think and behave! It is, however, the idea of the 'objective class' which is interesting since it permits the generation of flexible, rather than rigid categories of people. In particular we can see these objective classes as taking account of different concentrations of cultural and economic capital. The habitus, then, is the practical outcome of the class condition and will, for example, explain why people with the same amounts of wealth spend it in very different ways. It becomes interesting as a way of thinking about the development of gentrified inner-city areas which are shunned by those who prefer orderly suburbs; the phenomenon of post-tourism, which is not understood by those who seek holidays that are 'home plus'.

Habitus has mostly been used by geographers in the debate about the significance of human agents in the context of structuring (structuration theory) for it neatly (too neatly?) translates into the local configuration of practice.
In Bourdieu's equation $\{(\text{habitus})\ (\text{capital})\} + \text{field} = \text{practice}$ (p.101).

increasingly become the preserve of those who, either full or part time, buy into the rural idyll. The number of agricultural workers has dramatically declined throughout this century whilst the number of service-sector workers (and their families) and affluent retired is increasing. Many English villages in particular, have become self-consciously quaint and traditionalistic (this is not the same as being traditional) as people with urban-generated wealth assume pseudo-rural lifestyles which quickly become the new authenticity. Thrift (1989) is particularly percep-

BOX BB Cultural capital

Bourdieu's concept of cultural capital assumes that, for certain élite groups, culture is not just something which emerges from day-to-day practice but is instead consciously invested in and accumulated. Distinction resides in knowing what is 'good' in an aesthetic sense. Formal education can be an important part of cultural capital, but is only a part of the investment, which relies also on 'knowing' which sorts of music, literature, food, architecture, decor, clothing and leisure activities, etc., one 'ought' to prefer.

The acquisition of cultural capital can be formal, particularly through the *grandes écoles*, or informal, through family and association. It requires effort to learn to despise things which are easy to like, and to learn the codes of distinction. It necessitates a prioritization of aesthetics over utility or sensuality:

> The aesthetic disposition which tends to bracket off the nature and function of the object represented and to exclude any 'naive' reaction – horror at the horrible, desire for the desirable, pious reverence for the sacred – along with all purely ethical responses, in order to concentrate solely on the mode of representation, the style. . . . These conditions of existence, which are the precondition for all learning of legitimate culture, whether implicit and diffuse, as domestic cultural training generally is, or explicit as in scholastic training, are characterised by *the suspension and removal of economic necessity and by objective and subjective distance from practical urgencies, which is the basis of objective and subjective distance from groups subjected to those determinisms.* (Bourdieu, 1986, p.54; our emphasis)

Although scales of cultural capital exist in counterpoint to scales of economic capital (which implies the mere ability to buy), in practice the two value systems intertwine as the wealthy seek to improve their cultural standing, and cultural services become saleable, which is why one can talk of an investment in the acquisition of knowledge and the term 'capital' is appropriate.

The very notion of cultural capital is, as Bourdieu points out, arbitrary, exclusionary and élitist.

tive on the role of the rural for the new service class and the way in which it confers substance (what he calls 'bottom') to those whose claim to élite status is a little doubtful. This has been marked by a proliferation of rural lifestyle magazines and an increased circulation of the old 'country' journals such as *Country Life, The Field* and *Horse and Hound* which are bought as much because they are themselves seen as an ingredient of the 'traditional' lifestyle, as for their content which purports to mirror the concerns of authentic rural living. Possibly the most read pages of these journals are those carrying the advertisements for country properties and here we find that rural property prices reflect the successful commodification of the quality of country living.

Not all of those who would aspire to this lifestyle can, however, afford the housing costs or the commuting time, but as Urry (1989) points out, the countryside is also on sale for holidays and day trips. Tourism is a major source of rural income as pursuits like riding and shooting are commodified along with the products of revived craft industries, refurbished agricultural labourers' accommodation and imitation peasant food. The appearance of rural settlements changes as a prettified version of traditional life is constructed and villages slide into a fake version of some vaguely specified past – postmodern lifestyles dressed up in premodern costume. Rural shopping surprises those who have constructed their images only from television dramas. The baker, the butcher and the grocer have usually disappeared because people obtain their basics from suburban shopping complexes; in their place are fashion outlets, quality craft goods, antiques and restaurants, often run by women married to men employed in the higher echelons of the service sector. In Chapter 3 we dealt at some length with the commodification of the heritage site and the way in which this reflected back into environmental and house value. To some extent the whole of the rural is regarded as a part of the heritage and pseudo-pastoral lifestyles are purveyed throughout that part of the British countryside which is accessible to affluent urban areas.

Although most easily understood in the individual terms of housing and leisure, lifestyle considerations also need to be taken into account in the spatial practices of new commercial activity. The phenomenon of the business park, slower to catch on in Europe than in the United States, is particularly interesting here, as the rhetoric of exclusive living is appropriated in the selling of out-of-town commercial space. Self-consciously postmodern architecture, frequently mimicking the styles of grand country houses, is situated in landscaped parkland whilst 'villages', with shops and restaurants but no houses, serve as a focal point for the construction of a distinctive business community which is as self-creating as any new neighbourhood. Business parks as places are created out of space which would formerly not have been seen as appropriate. Certainly improvement in motorways and the spread of car-ownership played a big part in this development, as did the demand for large offices which could handle the new technology, but it is doubtful if business parks could ever have repaid the necessary investment if they had not been able to plug into the same lifestyle aspirations as those promoted by similar residential and leisure complexes. Moving onto a business park could be seen as a statement about being youthful and contemporary, offering one's high-quality workforce a superior environment and turning one's back on the old CBD or the out-of-date, shabby modernity of the peri-urban industrial estates. In the same vein, other urban services (i.e. cinemas, retail outlets and restaurants, etc.) started to situate themselves on greenfield sites close to motorway interchanges, almost invariably eschewing modernist styles.

Massey *et al.* have shown in their book *High-Tech Fantasies* (1992) that the task of image creation involved in the establishment of science parks was quite as important as any 'real' benefits of proximity to university-based intellectual capital. Siting a company on a science park identifies it with the sharp end of research and development, even though it turns out that companies on science parks do not actually 'tap in' to employ academics or sponsor research trials in university laboratories any more than similar companies situated elsewhere, though they are likely to use the university library, computer services, recreational and dining facilities.

The culture of a science park, based on the founding of an identifiable working community, was one of the major advantages to be bought with the location. Aggressively competitive male individualism, dedication to long hours and a willingness to travel long distances at short notice as well as an arrogance about one's intellectually élite status, all served to upgrade the quality, flexibility and productivity of the science park workforce. Managers talked about the way in which the ideal employee was a young bachelor, living with parents, a man with domestic support, but few responsibilities. The local culture of workaholism and high-performance cars made it difficult then for those with families, or for women, to fit in. Financial rewards could be high, but so was the value placed upon the exclusive working lifestyle and the agreement that work was the most important aspect of life. It is difficult to believe that such a strong and exclusive culture could have been built up in small companies not situated on a campus.

These commodifications of abstractions are rather different from those of factors of production and objects with use value. We have gone beyond exchange value into something closer to value pure and simple. Baudrillard's (1968) use of the notion of hyperreality in terms of the sign value of the commodity takes us some way to understanding how a high-performance car, or a particular way of dressing and deportment, can be part of the ethos whereby a science park, as a site, may become a desirable commercial location. In his more recent work Baudrillard (1993) has moved on to what he calls *viral* value. Whilst communicative (or *sign*) value refers back to structures of meaning relating into structures of power and control of wealth, and communication depends upon the ability to generate readable metaphors between realms of meaning, *viral* value has the potential for becoming completely astructural, depending upon metonym, contagion, contiguity and contamination. For Baudrillard the fractal rather than the structural is the means whereby one can begin to think about this new 'epidemic of value'. We shall try to capture some of the spirit of this sort of value (we hesitated about the word 'system', but it was obviously inappropriate) in the next chapter when we look at the spectacular, but in particular it can be contemplated in relation to the shift of the market for goods and services into the ether and away from the market*place*. The teleshopping which relies for its commercial 'environment' upon contextualizing programmes and advertising, the banks without branches, the telephone counselling services, the phone messages from Father Christmas, the horoscope and even the 'rub and shoot' lines, all atomize service into the *atmosphere* of commodification which then permeates into the supposedly private space of millions of homes, potentially worldwide.

13

Monuments and spectacles

Earlier in this book we looked at ways in which everyday places were constructed with respect to such notions as home, heritage, the exotic and the imaginary. We now wish to consider how many places are constructed first as monuments and then as broader, commodified spectacles. For thousands of years people, usually men, have created monuments which not only display values of wealth and individual, lineage and state power, but also serve in the construction of these. Obvious examples are the pyramids in Egypt, the Taj Mahal at Agra, the White House in Washington DC, the Hassan II mosque in Morocco and the Paris *Grands Projets*, all monuments to different forms of state power. Each of these has a practical use and an even greater signifying power, merging propagandist authority and celebration into a singular manifestation of the state, in their claim to eternity.

The Taj Mahal is not just a mausoleum erected by a king for his dead wife, it is physical ingredient in a love story of classic proportions (see Fig. 13.1). As a building it is so pure and exquisite that it seems that it cannot be corrupted by the tourist hype which pushes right up to its perimeter and, although one almost expects to feel disappointed by a monument so well-known, it defies any attempt to turn it into a cliché. Part of its power resides in the knowledge that it is an incomplete statement, that Shah Jahan had intended that it be matched by a black edifice containing his own remains but, having spent his declining years imprisoned, able to gaze out upon the Taj Mahal, he was buried there with Mumtaz. Not granted his own monument of kingly greatness, he was absorbed into his monument of enduring love. At the Taj Mahal, the sentiments of tragedy and aristocratic love become something one can experience, not simply contemplate. But the meaning of the Taj Mahal extends out beyond this, it becomes a symbol for India, for pre-imperial splendour and for the polyvalent attitudes to the Mogul Empire; it is a Muslim holy place in a predominantly Hindu society, but is espoused as an emblem beyond communal difference. Its name and image are used on

Fig. 13.1 The Taj Mahal

thousands of products, and hotels and restaurants run by Indians, Pakistanis and Bangladeshis worldwide bear its name.

Smaller monuments, such as Nelson's Column, or the Cenotaph in London, or the Arc de Triomphe in Paris, commemorate particular events, usually involving men's heroic deeds, particularly those concerning death or war. But they can also commemorate political milestones or resilience in the face of national or local disasters. Such monuments seem to be fixed points in urban landscapes, incorporating 'a "mysterious" sense of collective memory' (Harvey, 1989, p.85). They function as keystones in national mythologies; archetypal metaphors for the collective soul. The very term 'memorial' implies notions of a remembered past, but as we saw in Chapter 3, monuments open up poetic spaces of interpretation that promote particular futures and exclude others. But, like Ozymandias' statue, their message can become obscured or perverted (see Box CC). When political forces change rapidly, old memories can be violently erased and the toppling of monuments becomes a revolutionary act, symbolic of the toppling of a regime. Some monuments have extremely complicated pasts, used symbolically in a variety of ways. But although female figures may often be used as monumental representations of abstractions to be gazed at (see Fig. 13.2), embodying such values as Liberty and Justice, the story told is rarely other than heroic and masculine. Gregory's (1994) recent account of New York's Statue of

BOX CC *Ozymandias*

I met A traveller from an antique land
Who said: Two vast and trunkless legs of stone
Stand in the desert . . . Near them, on the sand.
Half sunk, a shattered visage lies, whose frown,
And wrinkled lip, and sneer of cold command,
Tell that its sculptor well those passions read
Which yet survive, stamped on these lifeless things,
The hand that mocked them, and the heart that fed:
And on the pedestal these words appear:
'My name is Ozymandias, king of kings;
Look on my works, ye Mighty, and despair!'
Nothing besides remains. Round the decay
Of that colossal wreck, boundless and bare
The lone and level sands stretch far away.

(Shelley)

Fig. 13.2 Jean-Baptiste Carpeaux's *Les Quatre Parties du Monde Soutenant la Sphere Celeste*

Liberty (drawing on Warner, 1985), for example, reveals its initial conception as an Orientalist symbol to light up the supposed darkness of the Middle East, before its shipment to the United States and its ironic, patriarchal portrayal of liberty as a woman.

Even the most seemingly innocent monuments and buildings are politically charged. Harvey's (1985) interpretation of the Basilica de Sacré Coeur in Paris, quickly destroys any romantic feelings one may have about the famous Paris landmark. Its massive presence commemorates the defeat of the Paris Commune uprising of 1871. Built on the site of the massacre of the martyrs of the commune, it is a monument to the triumph of bourgeois values, the ascendant power of right-wing clericalism and the installation of sentimental piety in opposition to the secular state. It is impossible to have an apolitical reaction to the Sacré Coeur once one knows its significance.

Monuments, then, are deliberate physical manifestations of ideology, a more or less massive inscription of triumphal or laudatory statements upon the landscape. There is nothing modest about what they proclaim. As we indicated in Chapter 3, Nietzsche (1980) and Heidegger (1962) – recognize a type of historiography which they call monumental – the telling of stories of great men and epic deeds, a grand romantic version of history, recalled with admiration for the heroes of the past. All physical monuments have the intention of inspiring awe through their great sweeps of history.

Tombs, as opposed to unmarked graves, are monuments, sometimes on a public scale, but frequently they are of no more than communal or family significance. They seem to contain and locate the essence of the person commemorated (and it is significant that a religion like Christianity should insist upon the emptiness of the tomb of its founder – a denial of the possibility of locating a transcendental deity). The tomb is a statement of continuity, giving the deceased an important place of influence after death and can often become a shrine or a pilgrimage centre, sometimes associated with miracles, other times with enhanced spirituality. But in Britain, with the increasing use of crematoria, tombs have become demonumentalized, whilst those cemeteries that do remain have assumed greater significance as heritage space.

So monuments may take many forms according to the message they are trying to convey. They may have a function related to their message, as in the case of a mausoleum; they may have no function other than their message, as in the case of a statue; or they may have a function which is different from their message, as in the case of Sacré Coeur. The latter type seems particularly fashionable today in the monumentalizing of state or commercial power. At first sight, it would seem that there is no escaping the vanity of the recent presidents of France as they seek to monumentalize themselves (Chaslin, 1985). But the Paris *Grands Projets* do have functions as opera houses, museums, libraries, offices and parks, although this would seem to be lost by the international media attention given to the architectural competitions for their design. Their mediatized inaugurations serve to reinforce the claim of Paris to be an international, not just a French, city.

Traditionally, the completion of monuments and buildings involves a spectacular opening ceremony, often employing carnivalesque elements of excess, celebration and enjoyment, as the notions of spectacle and monument become entwined. The spectacle encodes the heroic construction of the monument into the landscape. However, the creation of an urban spectacle is not always structured around a statue or a building, it can also take the more spatial expression of a park, a road or even that of a landscape poem or a painting. But, in whatever textual form, spectacular monuments are centred around issues of social control through excitation and enjoyment as well as discipline. As Harvey (1989, p.88) writes: '"Bread and circuses" is an ancient and well-tried formula for social control. It has frequently been consciously deployed to pacify restless or discontented elements in a population.' And politically, throughout history, the creation of licensed spectacles has involved the production and displacement of the subaltern other, through the maintenance of specific nationalisms.

As we have seen, the construction of spectacular, politically charged monuments has had a tendency to be specifically masculine in conception, privileging male vision and voyeurism. They are heroic not only in that they glorify great men but also in the sense that they are often places built on a masculine scale, from a masculine perspective, by men, for men, as their spaces of play, production, consumption and desire. The proliferation in the nineteenth century of tall monuments and offices, could be interpreted as phallic projections from those men who held state or commercial power. But they are not *simply* phallic symbols with psychosexual explanations, nor summits that we can walk away from (Morris, 1992), although we may be forgiven for thinking so if we were to read only Soja's account of *Postmodern Geographies* (1989). He admits that he prefers 'the monumental twenty-eight storey City Hall [Los Angeles], up to the 1920s the only erection in the entire region allowed to surpass the allegedly earthquake-proofing 150 foot height limitation' (p.224).

The expanding production of such monumental buildings in the Western world in the late nineteenth and early twentieth centuries was organized according to new consumption practices that were in turn linked to the greater social fluidity and economic dynamism caused by the industrial revolution (Glennie and Thrift, 1992). An increase in the pace of life occurred as traditional spatial barriers were overcome due to new transport technologies. The world seemed to collapse inwards as the time taken to physically traverse space was reduced (Harvey, 1989). Moreover, the way we represented this to ourselves also became faster, sowing the seeds of mass advertising. New print technology in Europe had produced newspapers that allowed people to realize that their lives ran alongside others, living many miles away, who were, at least superficially, similar to themselves.

Spectacular *events* began to globalize these new consumption practices. The series of World Industrial and Trade Exhibitions in London, Paris and Chicago, sought to reaffirm the identity of place in the midst of growing abstractions of space, often through the building of commercial monuments, such as Crystal Palace in London, or the Eiffel Tower in Paris. At the same time, many European cities were opened up from the inside with urban-clearance programmes, rewriting the city in modern form. Haussmanns' boulevards, for example, were not just magnificent linear statements cut across Paris, they were the means whereby the modern urban lifestyle was realized. They were part of the same movement which produced the arcades, the great department stores, shopping as a pastime and the (in)famous *flaneur*. We can see all of these as an international spectacularization of the commodity and commoditized relations. De Cauter (1993) has likened the early exhibitions to the growth in popularity of the panorama as a visual delight: 'As the panopticon marks the rise of the "disciplinary society" (Foucault), the panorama, as the first mass medium, marks the dawn of the "spectacular society" (Debord).' The point being that in a panorama the viewer 'takes enjoyment from a distant reality that can be possessed, that is always on the verge of being annexed or colonized' (p.3). 'The most urgent reason for setting up these enterprises seems to be the desire to visualise progress itself' (p.6).

The monuments and buildings built for these world exhibitions began to be overcoded by the pace of the spectacular events that surrounded them. The Eiffel Tower, for example, began to lose its primary meaning as an industrial symbol for the 1889 World Trade Fair. It quickly became the pure signifier to which a multitude of romantic meanings could be attached (Barthes, 1982; see Fig. 13.3). Like many tall buildings and monuments, it is both a monument to be seen, and to see from, and provided the city's tourists with an endless chain of metaphors (Duncan and Duncan, 1992). The romantically commodified and spectacular Eiffel Tower became used in thousands of advertisements for Paris, for France and for a range of supposedly sophisticated products, such as Yves St Laurent's perfume, *Paris*.

The Universal Exhibitions also displayed newly invented means of communicating, such as the camera – a way of viewing space through capturing time – the telephone and the wireless-telegraph. The cinema, the bicycle and the automobile were also all invented in the nineteenth century, and established the material foundations for later time-space compression beginning around the time of the First World War. Whilst these inventions changed our concepts of distance, their industrial mass production served to reinforce the centralized planning power of the city, exemplified by architects such as Le Corbusier. Dwellings were designed as monuments to urban modernity and as such became 'machines to live in' (Le Corbusier, 1927). After the Second World War, urban planning began to work with the new technologies to produce a distinctly modernist experience of space and time, incorporating Utopian views of highly structured schemes for living (Hall, 1988). But these transformations were by no means confined to Europe and America, Le Corbusier's vision of Chandigarh as the joint capital of Punjab, Harayana and Himachal Pradesh was more a monument to modernism than an embodiment of local urban practice.

In the late twentieth century cities continue to spectacularize themselves, competing through mediatized events such as Olympic games, Expos and cultural festivals. At contemporary Expos the pavilions of the contributing nations reflect their global economic positions. Whilst poorer countries exhibit industrial goods or folk art, the most spectacular displays of wealthy nations are no longer of products of national tradition, but of electronic performance. As Bennett (1991) indicates in his deconstruction of *Rainbowsphere*, the focus of the Australian pavilion at the Brisbane Expo of 1988, the ability to dazzle has taken over from material achievement. This audiovisual display was a high-technology representation of Aboriginal Dreamtime, which merged aboriginal history into postmodernity. This new spectacularization should be seen as 'intimately linked with the disappearance of a coherent representational system' (De Cauter, 1993, pp.20–1).

Cities now generate more of a sense of confusion as we are faced with an endless array of commodified signs and styles, metaphors and images, in both private and public places (Harvey, 1987, 1989). Whilst modernity had a penchant for monumentality in order to communicate a sense of authority, in moving from modernity to postmodernity some places have become *hyperreal*, losing much of their referentiality and meaning. Through computers and instantaneous telecommunications, today's spectacles are liable to be spatially diffuse, preventing the attachment of definite meanings, and producing a series of spatial *simulacra*; mediatized, uninterrupted circuits that refer back to no reality except the hyperreality of the image. In short, new communication systems, operating via satellites, seem to have changed our sense of space once again. Increasingly, the speed of information flows between individuals and between cities has accelerated to a point where 'Speed distance obliterates the notion of physical dimension. Speed suddenly becomes a primal dimension that defies all temporal and physical dimensions' (Virilio, 1991, p.18). But, as Harvey (1989, pp.293–4) has argued, 'The collapse of spatial barriers does not mean that the significance of space is decreasing'. On the contrary, smaller and smaller spatial differences take on increasing power, and 'as spatial barriers diminish so capitalism becomes much more sensitised as to what the world's spaces contain' (Harvey, 1989, p.294). This increases the volatility of finance capital flows around the world, destabilizing companies, cities and many Third World countries (Thrift, 1989b, 1993).

The French urban theorist, Paul Virilio, has termed today's cities 'overexposed', due to processes of 'dis-urbanization' that merge traditional dichotomies such as urban–rural, centre–periphery into a series of *interfaces* operating at different *speeds*. What he is saying is that the forces of time-space compression do not operate uniformly, and we experience, interact, or as he puts it, interface these forces through different technologies, such as the computer, the video, the supersonic jet, etc. We now enter the city electronically, either through computer *E-mail* or through the electronically guided fly-by-wire jet, and the city or *metropolis* has now, in a sense, extended itself to encompass the whole world. This information experience has transformed many people's relationships with space and time. As Virilio (1991, p.13) puts it: 'From here on, people can't be separated by physical obstacles or by temporal distances. With the interfacing of computer terminals and video monitors, distinctions of *here* and *there* no longer mean anything'. Architects such as Peter Wilson have begun to use Paul Virilio's ideas directly. Wilson has applied Virilio's idea of interfaces in his designs for Cosmos Street in Tokyo (see Fig. 13.3).

Virilio chooses the metaphor 'overexposed' to describe the city because the speed of telecommunications is, of course, the speed of light, and, just as an image becomes distorted when you overexpose a camera film by letting in too much light, Virilio is arguing that through the speed of telecommunications the city image is becoming distorted and transparent, lacking in dimensions. All that is left for reference is a series of points or *punctums*.

The use of transparent or reflective materials to allow light in also signifies the importance of the speed of light, of instantaneous information transmission. The buildings that are designed and built these days construct a postmodern sense of illusion by recoding functional spatial ideas into a celebration of display. Buildings are stripped of their centres and given masses of circulatory spaces. At La Défense, in Paris, we can see a little of what Virilio is referring to – offices made of glass, and a contortion of line and dimension. Clearly, the Grande Arche, designed by Spreckelsen and made of white marble, is a monument to light and transparency, as you see straight through its cubic dimensions. The Grande Arche is there, of course, to attract both capital and visitors, and is quickly catching up on the Eiffel Tower as the privileged icon of Parisian romantic power (Glancey, 1993), despite its primary symbolic intention to mirror the Arc de Triomphe and continue the historic axis of the Champs Elysées westwards out of Paris. An arch is a special kind of monument for it symbolizes continuity rather than boundary. Other

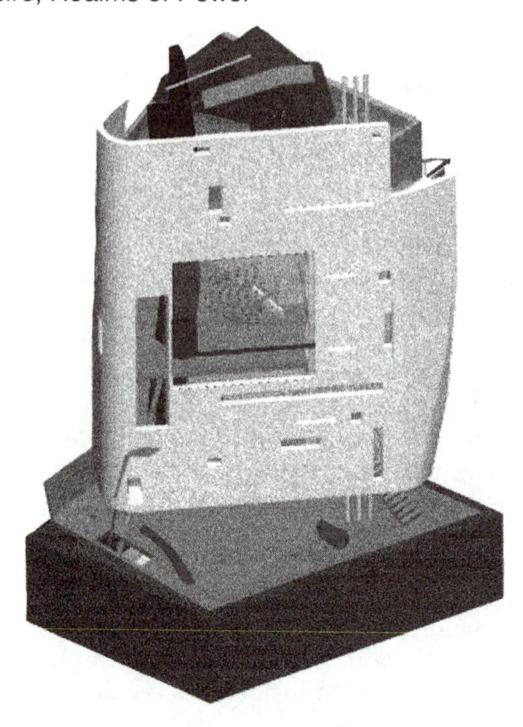

Fig. 13.3 Simulacrum of the interface on Cosmos Street, Tokyo

Parisian *Grands Projets* (see Fig. 13.4) that testify to the notion of transparency include Jean Nouvel's *Centre du Monde Arabe* which works 'intelligently' with light, I. M. Pei's glass pyramid at the Louvre and Dominique Perrault's new Bibliotheque de France, whilst the next project is for a *Tour sans fin* by Nouvel which will give the illusion of merging into the sky itself.

Just as the cultural theorist Walter Benjamin saw Paris' late-ninteenth-century shopping arcades as the defining feature of that decade's consumption practices, the French sociologist Jean Baudrillard sees the proliferation of Parisian *Grands Projets* as the defining spaces, or as he puts it, *monsters*, of the late twentieth century. Baudrillard (1982) has argued that places such as the Pompidou Centre (otherwise known as the *Beaubourg*) operate through a process of time-space implosion, or compression. In direct contrast to its architectural facade of rhizomic pipes and circuits that makes it look as if it has just exploded, it seems to suck energy and people into its vacuum via its external escalator (Gane, 1991; see Fig. 13.5). The area represents the postmodern articulation of continual development and experimentation, rather than the modern penchant for static monumentality, even though the intended physical flexibility of the building has been compromised to a rigidification of function. Inside the Pompidou Centre, despite its many uses as a Public

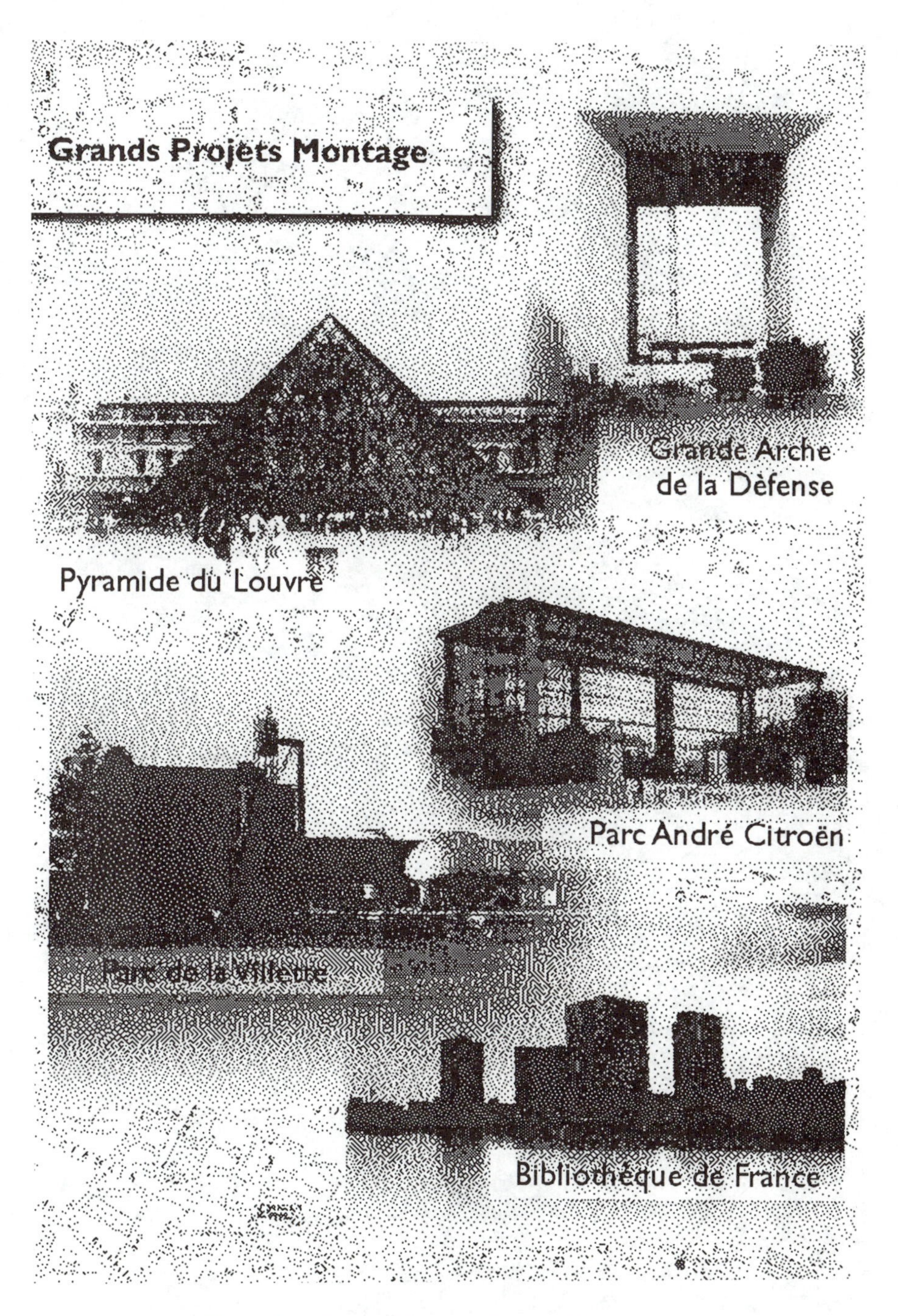

Fig. 13.4 Paris' *Grands Projets* Montage

Matisse
2151803345

Information Library, National Museum of Modern Art, and Centre for Industrial Creation, it operates as a number of controlled spaces 'in which you are both forced to circulate and yet deprived of orientation' (Heinich, 1988, p.209). Gane (1991, p.144) has pointed out that 'like a hypermarket it has no memory (unlike a traditional museum), only stock'.

The Parc de la Villette, designed largely by Bernard Tschumi has been called, somewhat perversely, the next logical step after the Beaubourg. It is a flat landscape dotted with functionally mutating follies, and is predicated on a denial of coherent meaning. There is no obvious edge to the park nor formal structure as in the case of the great Paris parks like the Jardin du Luxemburg. And, just like the area surrounding the Pompidou Centre, the Parc de la Villette's multiple entrances, juxtaposition of different routes and diverse connections allow the space to function at different speeds as we move in between the follies. To get to any point in the landscape we can take a number of different routes, each of which will get us to our destination, but at different times. In many of the initial designs for La Villette (particularly those submitted by the Office of Metropolitan Architecture under Rem Koolhaas), there was an explicit espousal of the philosophy of Deleuze and Guattari. It has also been said that La Villette represents one of the first attempts of architects to literally represent deconstruction. It is in this context that Baudrillard argues that these spectacular spatial monuments testify to the city's disorganization. They function as interchanges, places of 'expulsion, extradition, and urban ecstasy,' (Baudrillard, 1990a, p.105). Today, excitement is often felt in the trip to the monumental monster, rather than in monument itself.

We do not passively consume these new cultural spaces, though. Some individuals carefully plan their route and avoid the principle of free circulation. Most visitors are well-educated and are investing their desires for certain sociable and symbolic forms of cultural capital. Like Shields' (1991a) post-shoppers we may be simultaneously accepting and subverting the meaning of the paintings, books, sculptures, goods and services on offer in the shops and museums. Bourdieu's study of visitors to the Pompidou Centre revealed that 85 per cent of visitors had stayed at school until at least 18, more than 50 per cent were students or graduates, over 75 per cent were under 35, and 60 per cent came from the Paris region (Heinich, 1988). The majority of visitors are therefore young, local, middle class and intelligent, part of what Lash and Urry (1987) call the new petite bourgeoisie. Significantly, the majority of those who come to just circulate, wander and watch the street theatre of juggling acts are European tourists.

It is becoming clear that many urban spectacles are no longer defined by discrete monuments with definable meanings. Spectacles have become ecstatic or catastrophic events that no longer rely solely on a tangible monument, but on the television, cinema or computer screen. In spatial terms this implodes the separation between the purely imaginary experience of the spectacular place and the hyperreality of the surrounding area (Baudrillard, 1988). Disneyland, for example, 'is presented as an imaginary in order to make us believe that the rest is real, when in fact all of Los Angeles and the America surrounding it are no longer real but of the same order of the hyper-real and of simulation' (Baudrillard, 1988, p.17).

Baudrillard is not arguing that Los Angeles and America aren't real in a tangible sense, but in their senses of place and dwelling. Los Angeles may be one of the most 'superprofitable industrial growth poles in the world economy' (Soja, 1989, p.191), but it is also possibly the only place in the world that displays all of the extraordinary heterogeneity of the city simulacrum. In Los Angeles the use of space is asserted through the speed of telecommunications. World cities are at least verging on this simulation when their *imagescapes* become reterritorialized and deterritorialized continually and instantaneously by rioters, demonstrators, shoppers and police at a variety of different scales and speeds. We experience this simulation through Ted Turner's round the clock Cable News Network (CNN), which transforms our 'living space into a kind of global broadcast studio for world events' (Virilio, 1991).

When the 1992 south-central Los Angeles riots were televised as they happened, they became simultaneously part of fiction as well as fact. Like a great number of cultural events, in many ways the riots were carefully *timed* by subaltern participants for the media. It is not a case of blaming the media for inciting the violence, but it became clear that those involved waited until the cameras were in place before escalating the riots. People began to take part in the riots because they knew that they and the 'nowhere' suburban places where they lived would be 'captured' on television. 'INFORMATION, PERCEPTION, ACCELERATION. ... Don't forget that the LA riots were triggered, in part, by a home video tape pollinating its local seed of urban discontent across a global datascape before anyone in authority had time to react' (Beard, 1993, p.82). Similar processes happened recently in Tower Hamlets, east London, where a confrontation between the fascist British National Party and the Anti-Nazi League took place following the British National Party winning a seat on the local council. The confrontation erupted into a riot in front of television crews. As we are not made aware of precisely how the violence began, it quickly loses its referentiality and becomes just another conflict within a simulacrum.

McLuhan (1964, pp.3–4) once argued that 'During the mechanical ages we had extended our bodies in space. Today, after more than a century of electronic technology, we have extended our central nervous system itself in a global embrace'. Quite clearly, through the increased use of computers in societies, cities and bodies are beginning to enter circuits of computer-aided simulation. In virtual reality games, such as *Laserquest*, we place our bodies in a new spatial dimension of the simulacrum, but a dimension that we may still overcode in the way that we play the game. At a larger scale, through events such as pop musician's 'World Tours', a spectacularization of the individual body has taken place as artists project their identities onto landscapes. City landscapes become *facialized* by the performer (Deleuze and Guattari, 1988; Morris, 1992). Jean-Michel Jarre's transformation of both London's Docklands and Houston, Texas, through his musical and laser extravaganzas, were clear examples of how an individual's use of images and sound can literally inscribe itself on a place. But the individual can do this only with corporate backing, just as it becomes impossible for the individual to break into the global popular music scene without the ability to make the capital investment needed to spectacularize music through the use of video, as music ceases to be just a matter of sound.

The very idea of local concerts, as places of formal recitation, have been challenged by contemporary, large, improvized, chaotic gigs and raves in which the audience or spectators participate through various practices, such as stage-diving. Increasingly our grand performance spaces are being transformed as there are fewer stadiums and concert halls due to the 'global-vision broadcasting'. Meanwhile at mass staged events those physically present watch the same images on monitors as those watching television; they listen to amplified sound and are there to perform as audience as much as to be spectators.

In athletics, images of cities and images of bodies become interwoven in a simulacrum:

> As ticket prices and profits have given way to the exhorbitant fees the networks are willing to pay for exclusive broadcast rights, we have arrived at a point in which some are seriously contemplating completely eliminating the crowds from all major sporting events, and simply televising the matches and games in empty stadiums, filled with nothing but advertising billboards. (Virilio, 1991, p.91)

Like many pop stars, the bulging muscles of some athletes are pumped full of drugs, and, through training, diet and psychology their bodies lose their referentiality to 'real' bodies and become simulations, undermining their value in both moral and financial terms (amateur athletics is now big business after all). Similarly, but at yet another scale, the many computer games of the Olympics simulate the accelerating body of the athlete as he (we have yet to see a game with female figures) moves from Olympic city to Olympic city. The individualized body of both the city and the athlete are placed under commercial sponsporship, pharmocological or computer control and no longer have a definable form, but become 'a complex relation between differential speeds, between a slowing and an acceleration of particles' (Deleuze, 1981, p.165). Of course parts of our bodies aren't travelling literally at different speeds, but Deleuze and Guattari are trying to capture the feeling of chaos and uncertainty that we have in our everyday movement, particularly when we move from place to place at high speed on trains, planes and automobiles, or even across the striated space of an athletics track, a netball court or a rugby field, and try to co-ordinate our gendered and territorialized activities.

Clearly, we have shifted a long way from the idea of a simple monument to gaze at that represents the state or an individual. In the nineteenth century the notion of spectacle began to entwine the monument, but we now seem to be participating in a series of spatial simulacra that merge bodies and cities, spaces and places, landscapes and images, into what we can only term facialized monsters. Of course, all three heroic notions – monument, spectacle and simulacrum – can be found simultaneously in Western environments, but through the 'implosion of the media into the subject' we seem to be slipping more and more into the subtle and diffuse control of simulacra (Baudrillard, 1988).

This is not to say that we are somehow just docile armchair participants in these heroic environments. The notions of spectacle and simulacrum are predicated on individuals investing their desires into heroic landscapes, and we do this tactically. De Certeau (1984) has argued that spectacular, commodified spaces are always subtly practised through tactics of insinuation, presence and timing. Individuals subvert the dominant message of the spectacle sometimes through conscious decisions, sometimes through unconscious misreadings. As Shields (1989) has argued in the case of the West Edmonton Shopping Mall in Canada, where public space was transformed into carefully controlled private environment, the practice of 'spurious rebellions' and parodies by 'post-shoppers' who come to just look around, take their time, and not consume, points towards an inversion of the social conventions that normally surround shopping practices, a transgression of the no loitering dictum. At the Pompidou Centre, the wandering student might resist the official cultural spectacle inside, but may simultaneously take part in the broader simulacrum through purchasing kitsch souvenirs, or just hanging about in the carnival atmosphere outside. Similarly, Ley and Olds (1988) have argued that at the 1986 Vancouver World Trade Fair, by and large an élitist, white male, upper middle-class affair in its conception and planning, people engaged critically with the spectacle and very often used the event and its images actively not only as an educational resource and for their commodified fantasies, but also as an event for the reuniting of families and friends.

Monuments, spectacles and even simulacra can always be used as powerful political weapons and they may be essential aspects of revolutionary movements. But, what is increasingly certain is that, as processes of differential time-space compression operate, we enter more and more simulacra where the choice between false and real becomes, well, hyperreal.

SECTION IV

14

Doing cultural geography

Usually books end with a conclusion where the authors tell you what they have already said. We don't want to do this because we think that this sort of ending assumes that the people who wrote the book are confident in their opinions and want to go on drumming home their assurance about their authority. Instead, we believe that books, like everything else in this world, are based on positions that are changing, fluid, flexible and in a state of becoming – that nothing is ever a finished project, even though some things have to stop! This last chapter, then, is a line of flight.

Cultural geography is currently in a very exciting stage and many people believe that it is at the cutting edge not only of human geography, but the way that all geography is constituted; that if geography is to be anything other than the loose affiliation of the spatial aspects of the various social and physical sciences it is going to have to demonstrate the ways in which the geographical imagination constructs a dialogue between the various ways of studying and interacting with the environment. This is much more than the old 'man/land' relationship (whether one preferred to see the flow as from the physical to the human or the human to the physical in the process of construction); it is, instead, a matter of recognizing that we cannot see or interact with *anything* except through the medium of culture. But, as we have said many times, culture is itself not a given, definable object with boundaries and rigid structures in either time or space.

Much of cultural geography until the very recent past took a view of culture either as a given, where one would look at the *impact* of a culture in a particular setting, or as a product, where one would explain the cultural *form* emerging from a set of circumstances which were deemed to be prior to culture. The so-called 'new' cultural geography is disinclined to believe that culture has a form, or to assume that there is anything which is prior to culture – culture is everything we can experience. The economy is cultural, politics is cultural, artefacts are cultural, and nature is cultural. (If mountains, rivers and weather are not approached culturally there is a terrible lie uttered every time the word 'environment' is used!) To restrict cultural geography to such things as the study of the impact of religious belief on land use is to impoverish geography and to completely trivialize the vibrant, sinuous, uncatchable thing that is culture.

All of this means that cultural geography is very much up for grabs at the moment. Although some 'names' are better known than others and although the subject is attracting some of the brightest young minds in postgraduate geography, there really are no 'authorities'. It would be wonderful if things could stay this way, but we are world-weary enough to know that structures will emerge, new 'truths' will be laid down and, unless we are all very careful, the edge will be blunted and the radicalism evaporate in promotion struggles and the construction of *curriculum vitae*. For dull minds this will all be a reassuring development and will mark the settling down and respectability of the subject. Students will know what to 'swot up' for examinations, external examiners will no longer be faced with words they cannot find in their dictionaries as the whole thing becomes routinized. Then it will be time to migrate into some other realm, for it will not be the study of culture any more. Meanwhile, there is a considerable excitement within cultural geography and within geography about the cultural turn. Each year more students are taking the cultural route through their undergraduate studies and, where there is scope for independent studies, increasing numbers of people are submitting projects with a strong cultural element. We thought that it might be useful to end our book by looking at the sorts of things you might do.

Given that we just said that culture is everything, it would be nice if we could just say 'Do anything you like!' but the academic world has its own power structures, its own ways of deciding what is 'good' and if you are aiming at a high degree classification there is no point in committing suicide by being naive about your own local academic culture. Some of the things we suggest might not be appropriate in particular institutions, though there is no harm in establishing whether the local rules really are set in concrete or whether they just reflected current thinking at the time they were generated. For example, in many British universities there used to be a requirement that undergraduate dissertations generated and then proved/disproved a hypothesis, because this was thought to be good scientific practice. This was a very real block to cultural geographers who not only had to try to think up hypotheses which would not strangle what they were trying to study, but it actually prevented all work which was not positivist in orientation. It took only the very smallest push to get things changed, but it could not have been done without reference to the changes which were taking place within the field, particularly the philosophical shifts which were discrediting positivism. It would have been no use arguing simply on the grounds of interest or on the fact that some students did not like using quantitative methods.

Some departments demand that independent study be based on fieldwork or on the collection of original data, some still offer only the conventional written dissertation or essay as a means of assessment of independent work, whilst others, like Portsmouth, are willing to admit exhibitions, films and videos, audio productions, folios, posters, verbal presentations with a range of audio and visual aids, and various types of computer work, like hypertext, multimedia, interactive video and simulations. Many departments are now encouraging group work, with or without an individually assessed element and this, of course, allows far more ambitious projects to be undertaken (as well as forcing students to come to terms with their own and others' idiosyncrasies in areas where they thought there was only one 'of course' way of seeing things).

Obviously it is important to choose a project which fits the scale of the assessment to which it is directed. The conventional year-long 'dissertation or other independent study' of the British geography department anticipates *at least* 200 hours of work (the equivalent of five full-time working weeks), whereas an assessment built into a one-semester option course will allow something like 20 hours (this is often precisely where there is the most scope for team work). We do not want to provide recipes, but you will clearly need to see where you can build on or scale down, refine or roughen up our various suggestions. We would also hate to think that the only reason anyone might want to 'do' cultural geography was for an assessed project: we find that we do it all the time, for fun.

At the outset it is important to establish that work in cultural geography should not be undertaken for negative reasons – for example, because one does *not* want to use statistics, conduct a survey or draw maps. There have been times when cultural geography work has suffered from being seen as a soft option or drawing upon hobbyism. Experience at

Portsmouth has shown that cultural geography projects tend to be taken up by some of the best and the worst students, but not to attract the safe middle ground – the best know that they can find scope for their own creativity and interpretation within a theoretically taxing framework; the worst waffle on about what it feels like to live in their home village or what they think their favourite author was trying to say about nature. (That is the last we are going to say about the minimalist approach!)

If cultural geography projects are to have any validity it is essential that they are theoretically aware (and that 'theory' is not just tacked on cosmetically in order to demonstrate that one is up to date), for culture is about constituting meaning, i.e., theorizing. Although by no means all of the theory exists at the grand scale, we hope that we have conveyed a scepticism towards metatheorization in the preceding chapters, and shown that indigenous (or 'emic') theory can be as illuminating as the more conventionally academic explanatory context. We are not really saying any more than that a piece of research should have a *point* to it, it should not just introduce (potentially) interesting case studies and illustrations, it should argue with, refine, pull down or fit within structures of thought, according to your preferences. But it should always be clear where you are starting in terms of established theory, where you are going (even if the path is one you have to cut as you go along) and why you have chosen the material that you are using to make your point. If the reason you want to write a dissertation about a particular painter is simply that you 'like' her, we are afraid that you are going to have to spend some time theorizing that liking before starting anything useful. This is not really a problem, when we are subjectively drawn to a particular topic it is because it raises issues that interest us, but topics themselves are not inherently interesting though it is sometimes difficult to clarify the way in which one's subjective preferences are constituted. The answer frequently lies within one's autobiography, which will then suggest whether the interest arises in the realm of, say, gender relations, class politics or economics. What we are saying is, don't start a piece of research with the topic, start with what got the topic going in your mind and use this to construct a dialogue with what has been written not just on your topic, but with other things drawing upon a similar theoretical position. (So if you were looking at something like the construction of Englishness through an examination of the music of Elgar, there could be good reason to think about the different construction of Finnish identity in Sibelius and of Englishness in Kipling, as well as taking account of more abstract theoretical works on ethnic identity.)

It is not usually wise, within the time constraints of undergraduate research, to attempt to write a work of pure theory, without any grounding in what is conventionally called the real world (and if you are the sort of genius who could pull it off, it is unlikely that you would take any notice of what other people were recommending anyway). Given the short time you had been reading in the subject, it is likely that your work would be overly derivative and, given the sort of word limits set, that you would be unable to develop a sufficiently sophisticated position. However, this does not mean that one is locked into work in which the empirical outweighs the theoretical and in many good dissertations the bulk of the work has been done in the library rather than in the field.

With all these caveats, we will think about some of the sorts of subjects which seem manageable by undergraduates. Generally speaking we can (loosely) divide studies in cultural geography into the textual and the practical, where the textual tends to draw close to cultural studies and the practical to social anthropology. By 'textual' and 'practical' we are referring to the object of study, not the method of study; by this we mean focusing upon either texts or practices as objects of analysis, elaboration and interpretation, or deconstruction. (Historical work in cultural geography in particular can fall into both camps, for traces of the past, such as documents and buildings, can be read as text or taken as evidence of practice.)

It should be clear by now that when we refer to text we are following Ricoeur (1978) in the assumption that texts are not just written materials, they are anything that can be 'read', i.e., interpreted. (Yes, that can include social practice, but we are forced to draw lines sometimes, just for the sake of convenience!) There is a well-established tradition in cultural geography of taking artistic products – paintings, poems and novels, etc. – whether singly or as a corpus, school or genre and offering an interpretation in terms of the space, place or environmental construction contained within it. Cosgrove and Daniels, separately and together, have probably taken this endeavour further than anyone else in work which is aesthetically sensitive and politically aware. But textual work does not have to be restricted to 'high' art. There is plenty of scope for reading popular music (and not just the lyrics), film, advertising, TV, newspapers and magazines as ways of constructing geographical knowledge and, indeed, it is easy to argue that these have a greater impact upon the formation of ways of thinking and being in the world.

At the practical end of the field, ethnographic work currently reigns supreme with researchers participating in villages and neighbourhoods, workplaces and leisure activities, observing and absorbing the spatial and environmental practices and attitudes or the 'sense of place'. However, there is no reason why the study of cultural practice needs to be ethnographic (and if time is short, only limited and superficial work can be done by this method). Although neither of us is particularly drawn to the questionnaire-based survey, the quantitative methods which its use implies does not necessarily mean that the work is not cultural. We would caution against using questionnaires to assess people's attitudes rather than their practices; the quantification of the former is nearly always spurious and attitudes and values are best examined through the things people say and do in their own way, though this need not preclude interviewing.

We would like to outline some of the fields of study we have seen tackled over the last few years and show how they can be used to make interesting theoretical points.

Place advertising, whether for tourism, residence or business, is a rich vein for deconstruction and tends to be tackled by people who are interested in the commodification of place. We have seen studies of well-defined places where the projected image shifts through time and/or for different types of consumer but also studies which have taken a particular place use (like tourism or housing) and examined promotional variation across space. There is also scope for studies which expose residents to promotional material destined for outsiders in order to observe their reactions, whether cynical or approving. Burgess and Wood's (1989) study of the influence of advertising on business locational decision-making has been emulated at undergraduate level. It is quite important that these sorts of study do not make naive statements about the gap between advertising image and 'reality' as their main finding – it is obvious that advertisements are unreal, no one would be at all impressed by a promotion which did not promote! The clever thing is to show why particular unrealities work in particular contexts and which would be seen as insulting to the intelligence, irrelevant or incomprehensible. In doing this you would look at the symbolic elements employed when a particular place or use of generic places is niched or universalized.

Here you might consider whether the pictures used evoke moods, dramas, status positions, history, the present, nature or traditionalism; whether they contain people, if so what sort (you would be surprised how few old people appear in promotional material unless a specific service for the aged is being sold). The predominant colour used in certain types of advertising is worth exploring too, though you need to be aware of local colour symbolism and not read in your own assumptions – in a British context we find that a predominance of blue often indicates that there are aspirations of affluent tranquility, continuity and self-assuredness. You need to look at the particular objects used symbolically, for example in the late 1980s the windsurfer turned up in the most unlikely settings, though by the 1990s he was rather passé. He seemed to be used to evoke a range of ideas about getting ahead, being free, affluence, modernity, strength, virility, fun, competence and youth – it would be interesting to see which places are still buying into this encapsulation of the yuppie dream. The picture of the antique shop seems to be employed as a device to show gentility, solidity, 'character', heritage, aspiration and perhaps conservatism. The print styles and layout need considering for their messages about tone, mood and status. Then the actual words, not only for their overt messages but also style and register, use of particular words and phrases. All of these need to be considered both contextually and intertextually.

Intertextually, we need to think about the ways in which place-promotional materials 'talk to' each other and to other types of advertising, literature, film, TV, 'high' art, business 'presentations', report writing, comics, guide books, 'coffee table' books and so on in order to convey a range of polysemous meanings. Advertising finishes up not just selling place but also constructing ways of thinking about types of places, ways which ultimately filter back into both the geographical imagination and lifestyle and this is obviously an interesting subject for a dissertation.

Similar studies have been done where advertising materials are used to show how particular or generic places or environments are used to sell products – car advertising and the Alps, Highlands and 'mean streets' is an obvious one, as are ideas of nature in relation to a range of things from cosmetics to sliced bread – but there are interesting things to be done in terms of the placing of status, gender, ethnicity and age in order to invest products with character.

The whole range of media presentations of space, place, environment is also popular, generally building upon the approaches employed in Burgess and Gold's edited collection *Geography, the Media and Popular Culture* (1985). Here the theoretical orientation is often in terms of hegemony at different scales, for example local and national newspaper coverage, and regional, national and global broadcasting. It is interesting to see how geography students, often in dialogue with David Harvey (1989), have become increasingly involved in film studies and these are obviously particularly attractive when they are incorporated into video presentations rather than just deconstructed or analysed in written work. Film is inherently concerned with using space to convey meaning. The *mise en scène* is crucial to the power of filmic knowledge and is obviously spatial and ideological in that it constructs one's view. Media studies can also show philosophical shifts in the way the world is conceived of, or the way in which metaphors of place, space and environment are used. For example, structuralist and deconstructionist uses of spatial metaphor are very different.

So far we have only seen these sorts of studies as written dissertations, but they would work splendidly as multimedia productions, where, using a computer, one could combine text, music, static pictures, maps at various scales and fragments of video material. It becomes possible to scan a much larger number of images for comparison than one could incorporate into a bound dissertation. Multimedia would also permit numerous pathways through the material being presented. This really does not require enormous amounts of technical knowledge or equipment which is not available in most well equipped universities. There is also the relatively new phenomenon of the place promotion and souvenir video. We have yet to see this genre deconstructed, but it is certainly ripe for it. Although such a study could be done in the form of a linear text, it would be very amenable to a multimedia approach, too.

There is also a current vogue for studies, emerging from the work of Michel Foucault, on the issues of surveillance and control that we dealt with in Chapter 11. Philo's paper 'Foucault's geography' (1992) pointed the way for a new approach to the understanding of the micro-geographies not only of asylums, schools, workplaces and hospitals, but also of residential areas, retailing and recreational facilities. Such studies can be based on historical records or contemporary observation and use of planning documents. The point of such studies is to reveal the structures of power, whether rigid or sinuous, and the processes of connivance or opposition as they are revealed in the deployment of space. It becomes important to both literally and metaphorically map space and understand its practical and symbolic manipulation.

'Sense of place' studies have long been popular, particularly as group projects in the initial stages of cultural geography courses; we are uneasy about recommending them as a focus for a dissertation as they tend to operate at a rather low level of theory and to be more useful as an exercise in observation. Meinig's collection *The Interpretation of Ordinary Landscapes* and Relph's *Place and Placelessness* (1976) tend to be used as the template for these studies, but we would caution against too slavish a following since both are very thin on the concept of polysemy, tending to imply that there are privileged geographical 'readings' of landscapes, assuming that different meanings must come from different people with different points of view (the words polysemy, polyphony and multivocality are not synonymous). It seems that it is a little difficult to raise sense-of-place studies above the level of journalism, not that this is necessarily a criticism. Attractive presentations in this field usually draw upon a variety of sources from literature through interviews to personal 'feel'. They tend to be highly descriptive in nature, and the aim is richness of elaboration. We have a prejudice about attempting to classify places as authentic/inauthentic or value judgements about whether places have been 'spoiled', but there is scope for the approach to 'sense of place' which carefully accounts for the view of the observer.

Similar to sense-of-place studies are community studies. At their worst they are just an account of everything one can find out about what goes on within defined boundaries, usually of a village. We would suggest that if you are thinking about doing a community study, you need to start by problematizing first the notion of community and then the question of boundaries. It needs to be established that this particular strand of sociology became very unfashionable in the mid-1960s, but then was rediscovered with the rise of interpretive (often called postmodern) sociology and anthropology and is exemplified particularly in the work of Cohen (1982, 1986, 1987) and his followers. To oversimplify, this was to do with a shift from an assumption that a community was a more or less homogeneous group of people in place, to a view that people might use place as a means of constructing a (flexible) sense of belonging. Cohen's collection

Belonging (1982) is a useful introduction to the new community studies and can be read alongside Bell and Newby's (1971) critique of the genre prior to the shift. Community studies today are often conducted in order to throw light upon processes of transformation as new identities, accompanied by new structures of inclusion and exclusion, are locally invented in the face of wider economic and social change. So rural communities which see the function of their village changing from agriculture to tourism, commuter or retirement residence may well generate new ways of thinking about an insider/outsider divide and may start to think their rurality differently. We believe that community studies are best conducted according to the principles of the 'new' ethnography (which is not actually all that different from the best of the old ethnography). Turner and Bruner (1986), Clifford and Marcus (1986) and Marcus and Fischer (1986) are generally regarded as the best exponents of this, but the journal *Cultural Anthropology* generally gives useful illustrations of the method in palatable form. In brief, the strengths of this approach lie in an emphasis on catching the subjective meanings of members of the community being studied by allowing them a voice; the ethnographer then attempts to evoke, rather than analyse, the local reality. It is also considered important now that the ethnographer comes clean about his or her own position in the research, not pretending to be a detached observer, but explaining how she or he interacted with other people. This means being aware of one's own characteristics, personal as well as structural. An honest and engaging collection of studies on the role of gender in ethnographic fieldwork can be found in Bell *et al.* (1993) whilst Crawford and Turton's (1992) edited volume considers the theoretical issues in using film as a medium of ethnography.

For reasons we cannot fathom, there is often a mutual distrust between academic architects and geographers; this seems to us to be a cause for some regret because the different senses of space internalized in the two disciplines can enter into a useful dialogue. Studies of designed environments, whether pseudo-vernacular, themed or heroic, are well worth tackling for their insights into contemporary flows of desire, and awareness of architectural discourses is illuminating in terms of the compromises made between client desires and creative impulse. Work in this field can be theoretically oriented towards commodification, politics of spectacle, global/local interactions and lifestyle but also to ideas about gender, sexuality, ethnicity, disablement and age.

So far cultural geography has only begun to scratch the surface of the possibilities available to it. There is a very real need for work on construction of ideas about environmental crisis, risk and value, and the way in which these are mobilized in practice; the daring might like to look at the constitution of approved geographical knowledge in academic curricula. Although there has been a vogue for various types of ethnic and nationalist studies, there is still plenty of scope in this field, though we would always warn against doing such work as an 'outsider', unless the outsiderness itself is to be a part of the study. Informed, rather than simply disapproving, studies of processes of exclusion based upon race and ethnicity would be fascinating. We still have insufficient understanding of the implications of gender and age in the construction of place and the experience of space, and there is hardly anything of a geographical nature explicitly on masculinities. Aside from gender, sexuality and space is still very much in its infancy as a field of study (and quite a lot of work in this area seems to be desperate to demonstrate its 'acceptability' by mapping in very conventional ways). We consider that the work of Bell (1993) and Valentine (1993) stands out as an example of how one's sexuality may be sensitively foregrounded – and to those who would complain that sex has nothing to do with geography, we would counter that sex has everything to do with subjectivity and that pretending to be sexless has more to do with prudery than with academic respectability. We hardly know how to begin thinking in this area, though Colomina's collection *Sexuality and space* (1992) gives some fascinating pointers.

Much of the current interest in the cultural has emerged from a consciousness that transformations were taking place in the contemporary world, whereby the old explanatory frameworks were no longer convincing to enough people. At first we thought that this meant that a new economic stage implied that cultural changes were following along in their wake, then gradually we became conscious that the relationship between economy, politics and culture was more complex than we had ever realized. This means that practically the whole of conventional academic thought in all subjects, not just geography, needs to be overhauled – not simply because change has taken place, but because change has made us aware that knowledge has for too long been structured in favour of élite groups, whilst 'Others' (nearly everyone else in practice) are excluded into ignorance. The espousal of poststructuralist thinking (with the realization that all structures are contingent and that the recognition of flexible segmentations and lines of flight is just as important) means the possibility of empowering different types of thought which was previously castigated as illogical. We would like to feel that the greatest task for cultural geography is not the piling up of more and more case studies on topics which were once thought inappropriate, but a much more radical transgression, that of understanding and fusing into all of the illogics and anti-logics that inform our engagement with the world and the people within it.

Bibliography

Aaron, J. and Walby, S. (eds.) 1991: *Out of the Margins: Women's Studies in the Nineties*. London: The Falmer Press.

Aitken, S. and Zonn, L. 1993: Weir(d) sex: representation of gender-environment relations in Peter Weir's 'Picnic at Hanging Rock' and 'Gallipoli'. *Environment and Planning D: Society and Space* 11, 191–212.

Alvi, M. 1993: The double city. *Poetry Review* 83, 20.

Anderson, B. 1983: *Imagined Communities*. London: Verso.

Appadurai, A. 1990: Disjuncture and difference in the global cultural economy. *Theory, Culture and Society* 7, 295–310.

Appleby, S. 1990: Crawley: a space mythology. *New Formations* 11, 19–44.

Appleton, J. 1975: *The Experience of Landscape*. London: Wiley.

—— 1990: *The Symbolism of Habitat*. Seattle: University of Washington Press.

Ardener, E. 1987: 'Remote areas': some theoretical considerations. In Jackson, A. (ed.), *Anthropology at Home*. London: Tavistock.

Arensberg, C. and Kimball, S. 1940: *Family and Community in Ireland*. London: Peter Smith.

Armitage, S. 1992a: Cultural Studies. In *Kid*. London: Faber and Faber.

—— 1992b: *Xanadu: A Poem for Television*. Newcastle-upon-Tyne: Bloodaxe Books.

Asad, T. (ed.) 1973: *Anthropology and the Colonial Encounter*. London: Ithaca Press.

Attfield, J. 1989: Inside Pram Town: a case study of Harlow house interiors, 1951–61. In Attfield, J. and Kirkham, P. (eds.), *A View from the Interior: Feminism, Women and Design*. London: The Women's Press.

Atwood, M. 1987: *The Handmaid's Tale*. London: Virago.

Augé, M. 1992: *Non-Lieux: introduction à une anthropologie de la surmodernité*. Paris: Editions du Seuil.

Baggueley, P. and Mann, K. 1992: Idle, Thieving bastards? Scholarly representations of the 'underclass'. *Work, Employment and Society* 6, 113–26.

Bailey, F. 1971: *Gifts and Poison: the Politics of Reputation*. Oxford: Blackwell.

Balbus, I. 1987: Disciplining women: Michel Foucault and the power of feminist discourse. In Benhabib, S. and Cornell, D. (eds.), *Feminism as Critique*. Cambridge: Polity Press, 110–27.

Barnes, T. and Duncan, J. (eds.) 1992: *Writing Worlds: Discourse, Text and Metaphor in the Representation of Landscape*. London: Routledge.

Barrett, M. 1992: Words and things: materialism and method in contemporary feminist analysis. In Barrett, M. and Phillips, A. (eds.), *Destabilising Theory: Contemporary Feminist Debates*. Cambridge: Polity Press.

Barthes, R. 1982: The Eiffel Tower. In Sontag, S. (ed.), *A Barthes Reader*. London: Jonathan Cape.

Bartowski, F. 1989: *Feminist Utopias*. Lincoln N.E.: University of Nebraska Press.

Baudrillard, J. 1968: *Le System des objets*. Paris: Gallimard.

—— 1983: *Simulations*. New York: Semiotext(e).

—— 1988: *Selected Writings*. M. Poster (ed.). Cambridge: Polity Press.

—— 1989: *America*. London: Verso.

—— 1990: *Cool Memories*. London: Verso.

—— 1990: *Seduction*. Basingstoke: Macmillan.
—— 1990: *Revenge of the Crystal – Selected Writings on the Modern Object and its Destiny*. London: Pluto Press.
—— 1991a: The reality Gulf. *The Guardian*, 11 January, 25.
—— 1991b: *La Guerre du Golfe n'as pas eu lieu*. Paris: Galilee.
—— 1993: *The Transparency of Evil*. London: Verso.
Beard, S. 1993: Cyberville LA: an interview with William Gibson. *Arena* 41, 81–5.
Beck, U. 1993: *The Risk Society*. London: Sage.
Bell, C. and Newby, H. 1971: *Community Studies: An Introduction to the Sociology of the Local Community*. London: Allen and Unwin.
Bell, D. 1992: What we talk about when we talk about love: a comment on Hay (1991). *Area* 24, 409–10.
—— 1993: Citizenship and the politics of pleasure. Paper presented to the annual conference of the Institute of British Geographers.
Bell, D., Caplan, P. and Karim, W. (eds.) 1993: *Gendered Fields: Women, Men and Ethnography*. London: Routledge.
Benjamin, W. 1970: (Arendt, H., ed.) *Illuminations*. London: Jonathan Cape.
—— 1979: *One Way Street and Other Writings*. London: NLB.
—— 1989: *Paris, capitale du XIXe Siecle, le livre de passages*. Paris: Editions du Cerf.
Bennett, T. 1991: The shaping of things to come: Expo 88.
Bennington, G. 1989: Deconstruction is not what you think. In Papadakis, A., Cooke, C. and Benjamin, A. (eds.), *Deconstruction: Omnibus Volume*. London: Academy Editions.
Berger, P., Berger, B. and Kellner, H. 1973: *The Homeless Mind: Modernisation and Consciousness*. Harmondsworth: Penguin.
Berking, H. and Neckel, S. 1993: Urban marathon: the staging of individuality as an urban event. *Theory, Culture and Society* 10, 63–78.
Berman, M. 1982: *All that is solid melts into air: The experience of modernity*. London: Verso.
—— 1992: Why modernism still matters. In Friedman, J. and Lash, S. (eds.), *Modernity and Identity*. Oxford: Blackwell, 33–58.
Betjeman, J. 1958: Business girls. In *John Betjeman's Collected Poems*. London: John Murray.
Bhabha, H. 1988: The commitment to theory. *New Formations* 5, 5–23.
—— 1990: *Nation and Narration*. London: Routledge.
Binnie, J. 1992: Fucking among the ruins: postmodern sex in post-industrial places. Paper presented at the Conference of the Sexuality and Space Network. London: University College.
Bird, J. 1993: Dystopia on the Thames. In Bird, J., Curtis, B., Putnam, T., Robertson, G. and Tickner, L. (eds.), *Mapping the Futures: Local Cultures, Global Change*. London: Routledge.
Bishop, P. 1989: *The Myth of Shangri-La: Tibet, Travel Writing and the Western Creation of Sacred Landscape*. London: Athlone Press.
Bloch, M. 1971: *Placing the Dead: Tombs, Ancestral Villages and Kinship in Madagascar*. London: Seminar Press.
—— 1986: *From Blessing to Violence: History and Ideology in the Circumcision Ritual of the Merina of Madagascar*. Cambridge: Cambridge University Press.
Bloomfield, T. 1993: Resisting songs: negative dialectics in pop. *Popular Music* 12, 13–31.
Bogue, R. 1989: *Deleuze and Guattari*. London and New York: Routledge.
Bondi, L. 1991: Gender divisions and gentrification: a critique. *Transactions of the Institute of British Geographers* 16, 190–8.

—— 1992: Gender and dichotomy. *Progress in Human Geography* 16, 98–104.
Bondi, L. and Peake, L. 1988: Gender and the city: urban politics revisited. In Little, J., Peake, L. and Richardson, J. (eds.), *Women in Cities: Gender in the Urban Environment*. London: Hutchinson.
Bogarde, D. 1989: *An Orderly Man*. Harmondsworth: Penguin.
Bourdieu, P. 1971: *Outline of a Theory of Practice*. Cambridge: Cambridge University Press.
—— 1986: *Distinction: A Social Critique of the Judgement of Taste*. London: Routledge.
Boyd, K. 1991: Knowing your place: the tensions of manliness in boys' story papers 1918–1939. In Roper, M. and Tosh, J. (eds.), *Manful Assertions: Masculinities in Britain since 1800*. London: Routledge, 145–67.
Brah, A. 1991: Questions of difference and international feminism. In Aaron, J. and Walby, S. (eds.), *Out of the Margins: Women's Studies in the Nineties*. London and New York: The Falmer Press.
Brock, W. 1949: An account of 'Being and time', in Heidegger, M., *Existence and Being*. London: Vision Press.
Brönte, C. 1966: *Jane Eyre*. Harmondsworth: Penguin.
Burgess, J. and Gold, J. 1985: *Geography, the Media and Popular Culture*. London: Croom Helm.
Burgess, J. and Wood, P. 1988: Decoding Docklands: place advertising and decision-making strategies of the small firm. In Eyles, J. and Smith, D. (eds.), *Qualitative Methods in Human Geography*. Oxford: Polity Press, 94–117.
Burt, M. 1992: *Over the Edge: The Growth of Homelessness in the 1980s*. Washington DC: The Urban Institute Press.
Byron, M. 1993: Dish and cable take over: US television networks in the Caribbean. Paper presented at IBG Conference, Royal Holloway, University of London.
Callenbach, E. 1975: *Ecotopia*. New York: Bantam Books.
Campbell, J. 1964: *Honour, Family and Patronage*. Oxford: Oxford University Press.
Campbell, L. 1993: Anteros and intensity. In Broadhurst, J. (ed.), *Deleuze and the Transcendental Unconscious*. Warwick: PLI, 97–104.
Caplan, G. 1970: *The Elites of Barotseland 1878–1969: A Political History of Zambia's Western Province*. London: Hurst.
Cesara, M. 1982: *Reflections of a Woman Anthropologist: No Hiding Place*. New York: Academic Press.
Chapman, R. and Rutherford, J. (eds.) 1988: *Male Order: Unwrapping Masculinity*. London: Lawrence and Wishart.
Chaslin, F. 1985: *Les Paris de Francois Mitterand: histoire des grands projets architecturaux*. Paris: Gallimard.
Chopin, K. 1984: *The Awakening and Selected Stories*. Harmondsworth: Penguin.
Cixous, H. (tr. Levy, J.) 1985: *Angst*. London: John Calder.
—— 1986: *Dedans*. Paris: Des Femmes.
—— 1989: *L'heure de Clarice Lispector précédé de Vire l'orange*. Paris: Des Femmes.
—— (tr. Cornell, S. *et al.*) 1991a: *Coming to a Writing and Other Essays*. Cambridge MA and London: Harvard University Press.
—— (tr. Wing, B.) 1991b: *The Book of Promethea*. Lincoln N.E. and London: University of Nebraska Press.
Cixous, H. and Clement, C. (tr. Wing, B.) 1986: *The Newly Born Woman*. Minneapolis: University of Minnesota Press.
Clark, E. 1992: On blindness, centrepieces and complementarity in gentrification theory. *Transactions of the Institute of British Geographers* 17, 358–62.

Clifford, J. 1992: Travelling cultures. In Grossberg, L., Nelson, C. and Treichler, P. (eds.), *Cultural Studies*. London: Routledge.
Clifford, J. and Marcus, G. (eds.) 1986: *Writing Culture: The Poetics and Politics of Ethnography*. Berkeley: University of California Press.
Cloke, P., Philo, C. and Sadler, D. 1992: *Approaching Human Geography: An Introduction to Contemporary Theoretical Debates*. London: Paul Chapman.
Cockburn, C. 1983: *Brothers: Male Dominance and Technological Change*. London: Pluto Press.
Cohen, A. (ed.) 1982: *Belonging: Identity and Social Organisation in British Rural Cultures*. Manchester: Manchester University Press.
—— 1986: *Symbolising Boundaries: Identity and Diversity in British Cultures*. Manchester: Manchester University Press.
—— 1987: *Whalsay: Symbol, Segment and Boundary in a Shetland Island Community*. Manchester: Manchester University Press.
Coleman, A. 1985: *Utopia on Trial*. London: Hilary Shipman.
Colomina, B. 1992: The split wall. In Colomina, B. (ed.), *Sexuality and Space*. New York: Princeton Architectural Press.
Comaroff, J. 1985: *Body of Power, Spirit of Resistance: The Culture and History of a South African People*. Chicago and London: University of Chicago Press.
Connell, R. 1987: *Gender and Power: Society, the Person and Sexual Politics*. Cambridge: Polity Press.
Conrad, J. 1898 and 1975: Youth. In Conrad, J. (ed.), *Youth* and *The End of the Tether*. Harmondsworth: Penguin.
Coombes, A. 1992: Inventing the 'postcolonial': hybridity and constituency in contemporary curating. *New Formations* 18, 39–52.
Cooke, P. 1990: *Back to the Future*. London: Unwin Hyman.
Cope, J. and Krige, U. (eds.) 1968: *The Penguin Book of South African Verse*. Harmondsworth: Penguin.
Corden, A. 1992: Geographical development of the long-term care market for elderly people. *Transactions of the Institute of British Geographers* NS 17, 80–94.
Cosgrove, D. 1991: New world orders. In Philo, C. (compiler), *New Words, New Worlds: Reconceptualising Social and Cultural Geography*. Lampeter: IBG Social and Cultural Geography Study Group.
Craig, S. (ed.) 1992: *Men, Masculinity and the Media*. London: Sage.
Crang, M. 1994: Spacing times, telling times and narrating the past. Unpublished paper. *Time and Society*, forthcoming.
Crang, P. 1992: The politics of polyphony: reconfigurations of geographical authority. *Environment and Planning D: Society and Space* 10, 527–49.
Crawford, I. and Turton, D. (eds.) 1992: *Film as Ethnography*. Manchester: Manchester University Press.
Cresswell, T. 1993: Mobility as resistance, a geographical reading of Kerouac's 'On the Road'. *Transactions of the Institute of British Geographers* 18, 249–62.
Cross, M. and Keith, M. (eds.) *Racism, the City and the State*. London: Routledge.
D'Aguiar, F. 1985: *Mama Dot*. London: Chatto and Windus.
Dawson, G. 1991: The blond bedouin: Lawrence of Arabia, imperial adventure and the imagining of English-British masculinity. In Roper, M. and Tosh, J. (eds.), *Manful Assertions: Masculinities in Britain since 1800*. London: Routledge, 113–44.
Dear, M. 1988: The postmodern challenge: reconstructing human geography. *Transactions of the Institute of British Geographers* 13, 262–74.
Dear, M. and Wolch, J. 1987: *Landscapes of Despair: From Deinstitutionalisation to Homelessness*. Cambridge: Polity Press.
Debord, G. 1970: *Society of the Spectacle*. Detroit: Black and Red.

De Beauvoir. 1972: *Old Age*. Harmondsworth: Penguin.
De Cauter, L. 1993: The panoramic ecstasy: on world exhibitions and the disintegration of experience. *Theory, Culture and Society* 10, 1–24.
De Certeau, M. 1984: *The Practice of Everyday Life*. Berkeley: University of California Press.
Delaney, J. 1991: Ritual space in the Canadian museum of civilisation: consuming Canadian identity. In R. Shields (ed.), *Lifestyle Shopping: The Subject of Consumption*. London: Routledge, 136–49.
Deleuze, G. 1990: *The Logic of Sense*. New York: Columbia University Press.
—— 1991: Postscript on the societies of control. *October* 59, 3–8.
—— 1993a: *Critique et clinique*. Paris: Minuit.
—— 1993b: *The Fold: Leibniz and the Baroque*. London: Athlone.
Deleuze, G. and Guattari, F. 1980b: *On the Line*. New York: Semiotext(e).
Deleuze, G. and Guattari, F. 1983: *Anti-Oedipus: Capitalism and Schizophrenia*. London: Athlone.
Deleuze, G. and Guattari, F. 1988: *A Thousand Plateaus: Capitalism and Schizophrenia*. London: Athlone.
Denzin, N. 1991: *Images of Postmodern Society: Social Theory and Contemporary Cinema*. London: Sage Publications.
Derrida, J. 1981: *Dissemination*. Chicago: University of Chicago Press.
—— 1991: *The Derrida Reader: Between the Blinds*. London: Harvester Wheatsheaf.
Deutsche, R. 1991: Boys town: *Environment and Planning D: Society and Space* 9, 5–30.
Devi, M. 1993: Bayen. In Dharmarajan, G. (ed.), *Separate Journeys: Short Stories by Indian Women Writers*. London: Mantra Publishing Ltd.
Doane, M. A. 1987a: *The Desire to Desire: The Woman's Film of the 1940s*. Basingstoke: Macmillan.
—— 1987b: The 'woman's film': possession and address. In Gledhill, C. (ed.), *Home is where the heart is: Studies in melodrama and the woman's film*. London: BFI Books.
Doel, M. 1991: Installing deconstruction: striking out the postmodern. *Environment and Planning D: Society and Space* 9, 163–79.
—— 1994: Something resists: reading–deconstruction as ontological infestation (departures from the texts of Jacques Derrida). In Cloke, P., Doel, M., Matless, D., Philips, M. and Thrift, N. (eds.), *Writing Country*. London: Paul Chapman.
Donnison, D. 1992: Matter of life and death. *The Guardian*, 30 September.
Douglas, M. 1966: *Purity and Danger*. Harmondsworth: Penguin.
—— 1969: *Natural Symbols*. London: Barrie Cressett.
Douglas, M. and Wildavsky, A. 1983: *Risk and Culture: An Essay on the Selection of Technological and Environmental Dangers*. Berkeley: University of California Press.
Driver, F. 1985: Power, space and the body: a critical assessment of Foucault's 'Discipline and Punish'. *Environment and Planning D: Society and Space* 3, 425–46.
—— 1988: Moral geographies: Social science and the urban environment in mid-nineteenth century England. *Transactions of the Institute of British Geographers* 13, 275–87.
—— 1989: The historical geography of the work-house system in England and Wales, 1834–1883. *Journal of Historical Geography* 15, 269–86.
—— 1990: Discipline without frontiers? Representations of the Mettray reformatory colony in Britain, 1840–1880. *Journal of Historical Sociology* 3, 272–93.
—— 1992: Geography's empire: histories of geographical knowledge. *Environment and Planning D: Society and Space* 10, 23–40.

Duffy, C. 1993: *Never Go Back*. In *Mean Time*. London: Anvil.
Dumas, A. 1959: *Tangier to Tunis*. London: Brown Watson.
Du Maurier, D. 1981: *Rebecca*. London: Gollancz.
Duncan, J. 1980: The superorganic in American culture. *Annals of the Association of American Geographers* 70, 181–98.
—— (ed.) 1981a: *Housing and Identity: Cross Cultural Perspectives*. London: Croom Helm.
—— 1981b: From container of women to status symbol: the impact of social structure on the meaning of the house. In Duncan, J. (ed.): *Housing and Identity: Cross Cultural Perspectives*. London: Croom Helm.
—— 1993: Sites of representation: place, time and the discourse of the Other. In Duncan, J. and Ley, D. (eds.), *Place/Culture/Representation*. London: Routledge.
Duncan, J. and Duncan, N. 1992: Ideology and bliss: Roland Barthes and the secret histories of landscape. In Barnes, T. and Duncan, J. (eds.), *Writing Worlds: Discourse, Text and Metaphor in the Representation of Landscape*. London: Routledge.
Duncan, J. and Ley, D. (eds.) 1993: *Place/Culture/Representation*. London: Routledge.
Dunn, P. and Leeson, L. 1993: The art of change in Docklands. In Bird, J., Curtis, B., Putnam, T., Robertson, G. and Tickner, L. (eds.), *Mapping the Futures: Local Cultures, Global Change*. London: Routledge.
Duras, M. (tr. Bray, B.) 1985a: *The Lover*. London: Flamingo.
—— (tr. Bray, B.) 1985b: *The Sailor from Gibraltar*. London: John Calder.
—— (tr. Bray, B.) 1990: *Practicalities: Marguerite Duras Speaks to Jerome Beaujour*. London: Flamingo.
Durell, L. 1962: *The Alexandria Quartet*. London: Faber and Faber.
Duyves, M. 1992: Framing differences: inventing Amsterdam as a gay capital. Paper presented to the Conference of the Sexuality and Space Network. London: University College.
Eagleton, T. 1986: *Against the Grain: Essays 1975–1985*. London: Verso.
Eco, U. 1987: *Travels in Hyperreality*. London: Picador.
Eden, E. 1983: *Up the Country: Letters from India*. London: Virago.
Elam, Y. 1973: *The Social and Sexual Roles of Hima Women: A Study of Nomadic Cattle Breeders in Nyabushozi County, Ankole, Uganda*. Manchester: Manchester University Press.
Engels, F. 1892 and 1952: *The Condition of the Working-Class in England in 1844*. London: Allen and Unwin.
Enbe, C. 1989: *Bananas, Beaches and Bases: Making Feminist Sense of International Politics*. London: Pandora.
Eribon, D. 1989: *Michel Foucault, 1926–1984*. Paris: Flammarion.
Evans, J. 1993: First missionaries, then marketeers: Review of *The Oxford Book of Exploration*. *The Guardian*, 2 November, 8.
Eyles, J. 1985: *Senses of Place*. Warrington: Silverbrook Press.
Fainstein, N. 1993: Race, class and segregation: discourses about African Americans: *International Journal of Urban and Regional Research* 17, 384–404.
Fardon, R. (ed.) 1990: *Localizing Strategies: Regional Traditions of Ethnographic Writing*. Edinburgh: Scottish Academic Press; and Washington: Smithsonian Institution Press.
Featherstone, M. 1991: *Consumer Culture and Postmodernism*. London: Sage Publications.
—— 1993: *Gobal and Local Cultures*. In Bird, J., Curtis, B., Putnam, T., Robertson, G. and Tickner, L. (eds.), *Mapping the Futures: Local Cultures, Global Change*. London: Routledge.

Featherstone, M. and Hepworth, M. 1991: The mask of ageing and the post-modern life course. In Featherstone, M., Hepworth, M. and Turner, B. (eds.), *The Body: Social Process and Cultural Theory*. London: Sage Publications.
Featherstone, M., Hepworth, M. and Turner, B. (eds.) 1991: *The Body: Social Process and Cultural Theory*. London: Sage Publications.
Fitzgerald, F. 1986: *Cities on a Hill: A Journey Through Contemporary American Cultures*. New York: Simon & Schuster.
Fitzgerald, F. S. 1936: The Crack Up. In 1965: *The Stories of F. Scott Fitzgerald. Vol. 2. The Crack Up, with other Pieces and Stories*. Harmondsworth: Penguin Books.
Flaubert, G. 1973: *Madame Bovary*. Harmondsworth: Penguin.
—— *Correspondence. I. Janvier 1830 à Avril 1851*. Paris: Gallimard.
Fletcher, A. 1993: Fighting for the right to be different: *New Statesman and Society* 29 (10), 22–4.
Foster, G. 1976: *Tzintzuntzan: Mexican Peasants in a Changing World*. Boston: Little, Brown and Company.
Foucault, M. 1961: *Madness and Civilisation*. London: Tavistock.
—— 1972: *The Archaeology of Knowledge*. New York: Harper Colophon.
—— 1973: *The Birth of the Clinic*. Tavistock: London.
—— 1977: *Discipline and Punish: The Birth of the Prison*. London: Allen Lane.
—— 1979: *The History of Sexuality: Volume One, An Introduction*. London: Allen Lane.
—— 1984: Space, knowledge and power: an interview with Paul Rabinow. In Rabinow, P. (ed.), *The Foucault Reader*. Harmondsworth: Penguin.
—— 1986: *The History of Sexuality, Volume Three: The Care of the Self*. Harmondsworth: Penguin.
—— Theatrum philosophicum. In Bouchard, D. (ed.), *Language, Counter-Memory, Practice: Selected Essays and Interviews of Michel Foucault*. Ithaca: Cornell University Press.
Friedan, B. 1965 *The Feminine Mystique*. Harmondsworth: Penguin.
Fuss, D. 1989: *Essentially Speaking: Feminism, Nature and Difference*. London: Routledge.
—— 1992: Essentially speaking: Luce Irigaray's language of essence. In Frazer, N. and Bartky, S. (eds.), *Revaluing French Feminism: Critical Essays on Difference, Agency and Culture*. Bloomington and Indianapolis: Indiana University Press.
Gane, M. 1991: *Baudrillard's Bestiary: Baudrillard and Culture*. London: Routledge.
Gans, H. 1993: From 'underclass' to 'undercaste': some observations about the future of the postindustrial economy and its major victims. *International Journal of Urban and Regional Research* 17, 327–35.
Gardiner, M. 1987: *Footprints on Malekula: A Memoir of Bernard Deacon*. London: Free Association Books.
Geraghty, C. 1991: *Women and Soap Opera: A Study of Prime Time Soaps*. Cambridge: Polity Press.
Gilman, C. P. 1892, 1985: *The Yellow Wallpaper*. London: Virago.
—— 1979: *Herland*. New York: Pantheon Books.
Gilroy, P. 1993: *Small Acts: Thoughts on the Politics of Black Culture*. London: Serpent's Tail.
Ginsberg, A. 1963: The green automobile. In Ginsberg, A. (ed.), *Reality Sandwiches*. San Francisco: City Lights, 11–16.
Girouard, M. 1978: *Life in the English Country House*. Harmondsworth: Penguin.
Glancey, J. 1993: That old Paris embrace amid the new brutalism: *The Guardian*, 17 March, 22.

Glennie, P. and Thrift, N. 1992: Modernity, urbanism and modern consumption: *Environment and Planning D: Society and Space* 10, 423–43.
Godelier, M. 1986: *The Mental and the Material*. London: Verso.
Goetz, E. 1992: Land-use and homeless policy in Los Angeles. *International Journal for Urban and Regional Research* 16, 540–54.
Goffman, E. 1968: *Stigma: Notes on the Management of Spoiled Identity*. Harmondsworth: Penguin.
Gold, R. and Gold, M. 1990: 'A place of delightful prospects': promotional imagery and the selling of suburbia. In Zonn, L. (ed.), *Place Images in Media: Portrayal, Experience and Meaning*. Maryland: Bowman and Littlefield.
Golden, S. 1992: *The Women Outside: Meaning and Myth of Homelessness*. Berkeley: University of California Press.
Gould, P. 1991: Dynamic structures of geographic space. In Brunn, S. and Leinbach, T. (eds.), *Collapsing Space and Time: Geographic Aspects of Communication and Information*. London: Harper Collins Academic.
Gough, K. 1968: Anthropology: child of imperialism. *Monthly Review* 19, 11.
Gradidge, R. 1980: *Dream Houses: The Edwardian Ideal*. London: Constable.
Gramsci, A. 1971: *Selections From the Prison Notebooks*. London: Lawrence and Wishart.
Gregory, D. 1994: *Geographical Imaginations*. Oxford: Blackwell.
Grimble, A. 1952, 1981: *A Pattern of Islands*. Harmondsworth: Penguin.
Grosz, E. 1988: The in(ter)vention of feminist knowledges. In Caine, B., Grosz, E. and de Lepervanche, M. (eds.), *Crossing Boundaries: Feminism and the Critique of Knowledges*. Sydney: Allen and Unwin, 92–104.
Guha, R. and Spivak, G. 1988: *Selected Subaltern Studies*. New York: Oxford University Press.
Habermas, J. 1987: *The Philosophical Discourse of Modernity*. Cambridge: Polity Press.
Hall, P. 1988: *Cities of Tomorrow*. Oxford: Blackwell.
Hall, S. 1968: *The Hippies: An American 'Moment'*. Birmingham: University of Birmingham Press.
Hamnett, C. 1991: The blind men and the elephant: the explanation of gentrification. *Transactions of the Institute of British Geographers* 16, 173–89.
—— 1992: Gentrifiers or lemmings: a response to Neil Smith. *Transactions of the Institute of British Geographers* 17, 116–19.
Hanbury-Tenison, R. 1993: *The Oxford Book of Exploration*. Oxford: Oxford University Press.
Hannah, M. 1993: Foucault on theorizing specificity: *Environment and Planning D: Society and Space* 11, 349–63.
Hannertz, U. 1990: Cosmopolitans and locals in world culture. *Theory, Culture and Society* 7, 237–52.
Harley, J. 1990: Deconstructing the map. In Barnes, T. and Duncan, J. (eds.), *Writing Worlds: Discourse, Text and Metaphors in the Representation of Landscape*. London: Routledge, 231–47.
Harris, O. and Gow, P. 1985: British Museum's representation of Amazonian Indians. *Anthropology Today* 1.5.
Harrison, J. 1988: 'The spirit sings' and the future of anthropology. *Anthropology Today* 4.6.
Harrison, P. 1985: *Inside the Inner City: Life Under the Cutting Edge*. Harmondsworth: Penguin.
Hart, K. 1987: Commoditisation and the standard of living. In Sen, A. and Hawthorn, G. (eds.), *The Standard of Living*. Cambridge: Cambridge University Press.

Hartsock, N. 1990: Foucault on power: a theory for women? In Nicholson, L. (ed.), *Feminism/Postmodernism*. London: Routledge, 157–75.
Harvey, D. 1985: *Consciousness and the Urban Experience*. Oxford: Blackwell.
—— 1987: Flexible accumulation through urbanization: reflections on post-modernism in the American city. *Antipode* 19, 260–86.
—— 1989: *The Condition of Postmodernity*. Oxford: Blackwell.
—— 1993: From space to place and back again: reflections on the condition of postmodernity. In Bird, J., Curtis, B., Putnam, T., Robertson, G. and Tickner, L. (eds.), *Mapping the Futures: Local Cultures, Global Change*. London: Routledge.
Haug, W. 1987: Commodity aesthetics, ideology and culture. New York: International General.
Heath, A. 1992: The attitudes of the underclass. In Smith, D. J. (ed.), *Understanding the Underclass*. London: Policy Studies Institute.
Hebdige, D. 1979: *Subculture: The Meaning of Style*. London: Methuen.
—— 1990: Subjects in space. *New Formations* 11, v–x.
—— 1993: Redeeming witness: in the tracks of the homeless vehicle project. *Cultural Studies* 7, 173–223.
Heelas, P. and Morris, P. 1992: *The Values of the Enterprise Culture: The Moral Debate*. London: Routledge.
Heidegger, M. 1962: *Being and Time*. New York: Harper and Row.
—— 1966: *Discourse on Thinking*. New York: Harper and Row.
—— 1971: *Poetry, Language, Thought*. New York: Harper and Row.
—— 1980: *An Introduction to Metaphysics*. New Haven, CT: Yale University Press.
Heinich, N. 1988: The Pompidou Centre and its public: the limits of a utopian site. In Lumley, R. (ed.), *The Museum Time-Machine*. London: Routledge, 199–212.
Hélias, P. 1975: *Le Cheval d'orgueil: mémoires d'un Breton du pays Bigouden*. Évreux: Plon.
—— (tr. Guicharnaud, J.) 1978: *The Horse of Pride: Life in a Breton Village*. New Haven, CT: Yale University Press.
Heron, L. (ed.) 1985: *Truth, Dare or Promise: Girls Growing Up in the Fifties*. London: Virago.
—— (ed.) 1993: *Streets of Desire: Women's Fiction of the Twentieth Century City*. London: Virago.
Hershman, P. 1974: Hair, sex and dirt. *Man* 9, 274–98.
—— 1981: *Punjabi Kinship and Marriage*. Delhi: Hindustan Publishing Corporation.
Hetherington, K. 1991: Stonehenge and its festival: spaces of consumption. In Shields, R. (ed.), *Lifestyle Shopping: The Subject of Consumption*. London: Routledge.
Hewison, R. 1987: *The Heritage Business: Britain in Climate of Decline*. London: Methuen.
Hilton, J. 1933: *Lost Horizon*. London: Macmillan.
Hobsbawm, E. and Ranger, T. (eds.) 1983: *The Invention of Tradition*. Cambridge: Cambridge University Press.
Honneth, A. 1986: The fragmented world of symbolic forms: reflections on Pierre Bourdieu's sociology of culture. *Theory, Culture and Society* 3, 55–66.
hooks, b. 1991: *Yearning: Race, Gender and Cultural Politics*. London: Turnaround.
—— 1992: Representing whiteness in the black imagination. In Grossberg, L., Nelson, C. and Treichler, P. (eds.), *Cultural Studies*. London: Routledge.
Horner, A. and Zloznik, S. 1990: *Landscapes of Desire: Metaphors in Women's Fiction*. Hemel Hempstead: Harvester Wheatsheaf.

Houtman, G. 1985: Survival International: going public on Amazonian Indians. *Anthropology Today* 1.5.
Huxley, A. 1932: *Brave New World*. London: Chatto and Windus.
Ishihara, S. 1991: *The Japan that Can Say No: Why Japan Will Be First Among Equals*. New York: Simon & Schuster.
Irigaray, L. (tr. Hand, S.) 1987: Sexual difference. In Moi, T. (ed.), *French Feminist Thought*. Oxford: Basil Blackwell.
—— (tr. Collie, J. and Still, J.) 1992: *Elemental Passions*. London: Athlone.
Jackson, D. 1990: *Unmasking Masculinity: A Critical Autobiography*. London: Unwin Hyman.
Jackson, P. (ed.) 1987: *Race and Racism: Essays in Social Geography*. London: Routledge.
—— 1989a: Geography, race and racism. In Peet, R. and Thrift, N. (eds.), *New Models in Geography: Volume 2*. London: Unwin Hyman.
—— 1989b: *Maps of Meaning*. London: Unwin Hyman.
—— 1991: The cultural politics of masculinity: towards a social geography. *Transactions of the Institute of British Geographers* 16, 199–213.
Jameson, F. 1985: Postmodernism and consumer society. In Foster, H. (ed.), *Postmodern Culture*. London: Pluto Press, 111–25.
Jeater, D. 1992: Roast beef and reggae music: the passing of whiteness. *New Formations* 18, 107–21.
Jencks, C. 1977: *The Language of Post-Modern Architecture*. London: Academy editions.
Jenkins, R. 1992: Salvation for the fittest? A West African sportsman in the age of the new imperialism. In Mangan, J. (ed.), *The Cultural Bond, Sport, Empire, Society*. London: Frank Cass.
Jonker, I. 1968: I Don't Want Any More Visitors. In Cope, J. and Krige, U. (eds.), *The Penguin Book of South African Verse*. Harmondsworth: Penguin.
Jordan, T. and Rountree, L. 1982: *The Human Mosaic: A Thematic Introduction to Human Geography*, 3rd edition. New York: Harper and Row.
Katz, C. 1993: Growing girls/closing circles: limits on the spaces of knowing in rural Sudan and US cities. In Katz, C. and Monk, J. (eds.), *Full circles: Geography of Women over the Life Course*. London: Routledge.
Keats, J. 1973: *The Complete Poems*. Harmondsworth: Penguin.
Keith, M. and Cross, M. 1993: Racism and the postmodern city. In Cross, M. and Keith, M. (eds.), *Racism, the City and the State*. London: Routledge, 1–30.
Kerouac, J. 1957: *On the Road*. New York: Viking Press.
—— 1959: *Maggie Cassidy*. London: Panther.
—— 1962 and 1990: *Lonesome Traveller*. London: Paladin.
Kidder, T. 1993: The last place on earth. *Granta* 44, pp.9–48.
Kingsley, M. 1897 and 1982: *Travels in West Africa: Congo Francais, Corisco and Cameroons*. London: Macmillan.
Koegel, P. 1992: Through a different lens: an anthropological perspective on the homeless mentally ill. *Culture, Medicine and Psychiatry* 16, 1–22.
Koyabashi, A. and Mackenzie, S. (eds.) 1989: *Remaking Human Geography*. London: Unwin Hyman.
Kristeva, J. 1982: *Powers of Horror: An Essay on Abjection*. New York: Columbia University Press.
Kroeber, A. 1952: *The Nature of Culture*. Chicago: University of Chicago Press.
Kumar, K. 1987: *Utopia and Anti-Utopia in Modern Times*. Oxford: Basil Blackwell.
—— 1991: *Utopianism*. Milton Keynes: Open University Press.
Lacan, J. 1977: *Ecrits*. London: Tavistock.
—— 1982: The meaning of the phallus. In Mitchell, J. and Rose, J. (eds.). *Femi-*

nine Sexuality: Jacques Lacan and the Ecole Freudienne. London: Macmillan, 74–85.
Laermans, R. 1993: Learning to consume: early department stores and the shaping of the modern consumer culture (1860–1914). *Theory, Culture and Society* 10, 79–102.
La Fontaine, J. (ed.) 1978: *Age and Sex as Principles of Social Differentiation*. London: Academic Press.
Lan, D. 1985: *Guns and Rain: Guerrillas and Spirit Mediums in Zimbabwe*. Berkeley: University of California Press.
Lash, S. and Urry, J. 1987: *The End of Organised Capitalism*. Cambridge: Polity Press.
—— 1994: *Economies of Signs and Space*. Cambridge: Polity Press.
Lash, S., Urry, J. and Friedman, J. (eds.) 1993: *Modernity and Identity*. London: Blackwell.
Leach, E. 1978: *Culture and Communication: The Logic by which Symbols are Connected*. Cambridge: Cambridge University Press.
—— 1985: Amazonian Indians (letter). *Anthropology Today* 1.5.
Le Corbusier. 1927: *Towards a New Architecture*. London: The Architectural Press.
Le Guin, U. 1988: *Always Coming Home*. London: Grafton.
Lefebvre, H. 1991: *The Production of Space*. Oxford: Blackwell.
Lessing, D. 1953 and 1991: A home for the highland cattle. In *Five*. London: Palladin.
—— 1957 and 1968: *Going Home*. London: Grafton Books.
—— 1964: *Martha Quest*. London: Panther.
—— 1979: *Shikasta*. London: Jonathan Cape.
Levi-Strauss, C. 1962: *Le Totemisme aujourd'hui*. Paris: Presses Universitaires de France.
Lewis, C. S. 1964: *The Narnia Chronicles*. Harmondsworth: Puffin.
Lewis, G. 1985: From deepest Kilburn. In Heron, L. (ed.), *Truth, Dare or Promise: Girls Growing Up in the Fifties*. London: Virago.
Lewis, O. 1966: The culture of poverty. *Scientific American* 215, 19–25.
Lewis, P. 1991: Mummy, matron and the maids: feminine presence and absence in male institutions, 1934–63. In Roper, M. and Tosh, J. (eds.), *Manful Assertions: Masculinities in Britain since 1800*. London: Routledge.
Ley, D. 1989: Fragmentation, coherence and the limits to theory in human geography. In Koyabashi, A. and Mackenzie, S. (eds.), *Remaking Human Geography*. London: Unwin Hyman.
Ley, D. and Olds, K. 1988: Landscape as spectacle: world's fairs and the culture of heroic consumption. *Environment and Planning D: Society and Space* 6, 191–212.
Ley, D. and Samuels, M. (eds.) 1978: *Humanistic Geography: Prospects and Problems*. London: Croom Helm.
Light, A. 1991: *Forever England: Femininity, Literature and Conservatism Between the Wars*. London: Routledge.
Lispector, C. 1986: *The Hour of the Star*. Manchester: Carcanet Press.
Lively, P. 1987: *Moon Tiger*. Harmondsworth: Penguin.
Livingstone, D. 1992: *The Geographical Tradition*. Oxford: Blackwell.
Loudon, J. 1843: *On the laying out, planting and management of cemeteries and on the improvement of churchyards*. London.
Lowenthal, D. 1985: *The Past is a Foreign Country*. Cambridge: Cambridge University Press.
Lowry, M. 1933: *Ultramarine*. London: Jonathan Cape.
Lynch, K. 1972: *What Time is this Place?* London: MIT Press.

Lyotard, J.-F. 1984: *The Postmodern Condition, a Report on Knowledge*. Manchester: Manchester University Press.
Maccannell, D. 1992: *Empty Meeting Grounds: The Tourist Papers*. London: Routledge.
Mangan, J. (ed.) 1992: *The Cultural Bond, Sport, Empire, Society*. London: Frank Cass.
Mangan, J. and Walvin, J. (eds.) 1987: *Manliness and Morality: Middle-Class Masculinity in Britain and America 1800–1940*. Manchester: Manchester University Press.
Mann, K. 1992: *The Making of an English 'Underclass': The Social Dimensions of Welfare and Labour*. Milton Keynes: Open University Press.
Manning, O. 1981: *The Balkan Trilogy*. Harmondsworth: Penguin.
Marcus, G. 1992: Past, present and emergent identities: requirements for ethnographies of late twentieth century modernity worldwide. In Friedman, J. and Lash, S. (eds.), *Modernity and Identity*. Oxford: Blackwell, 309–30.
Marcus, G. and Fischer, M. 1986: *Anthropology as Cultural Critique: An Experimental Moment in the Human Sciences*. Chicago: University of Chicago Press.
Marcuse, P. 1993: What's so new about divided cities? *International Journal of Urban and Regional Research* 17, 355–65.
Marin, L. 1991: Frontieres, limites, limes: les recis de voyage dans L'Utopie de Thomas More. *Frontieres et limites: geopolitique, literature, philosophie*. Paris: Centre Georges Pompidou, 105–30.
Marx, K. 1974: *Capital*. London: Lawrence and Wishart.
Massey, D. 1991: Flexible sexism. *Environment and Planning D: Society and Space* 9, 31–57.
—— 1992: A place called home? *New Formations* 17, 3–15.
—— 1993: Power-geometry and a progressive sense of place. In Bird, J., Curtis, B., Putnam, T., Robertson, G. and Tickner, L. (eds.), *Mapping the Futures: Local Cultures, Global Change*. London: Routledge.
Massey, D., Quintas, P. and Wield, D. 1992: *High-Tech Fantasies: Science Parks in Society, Science and Space*. London: Routledge.
Massumi, B. 1992: *A User's Guide to Capitalism and Schizophrenia: Deviations from Deleuze and Guattari*. Cambridge, Mass.: MIT Press.
Matless, D. 1992: An occasion for geography: landscape, representation and Foucault's corpus. *Environment and Planning D: Society and Space* 10, 41–56.
Mattelart, A. 1991: *Advertising International: The Privatisation of Public Space*. London: Routledge.
Maugham, W. S. 1919: *The Moon and Sixpence*. London: Heinemann.
Mayle, P. 1990: *A Year in Provence*. London: Pan Books.
McCann, J. 1986: The social impact of drought in Ethiopia: oxen, households, and some implications for rehabilitation. In Glantz, D. (ed.), *Drought and Hunger in Africa*. Cambridge: Cambridge University Press.
McCracken, G. 1989: 'Homeyness': a cultural account of one constellation of consumer goods and meaning. In Hirschman, E. (ed.), *Interpretative Consumer Research*. Provo, UT: Association for Consumer Research.
McDonald, M. 1986a: Brittany: politics and women in a minority world. In Ridd, R. and Callaway, H. (eds.), *Caught Up in Conflict: Women's Responses to Political Strife*. Basingstoke: Macmillan.
—— 1986b: Celtic ethnic kinship and the problem of being English. *Current Anthropology* 27, 333–41.
McDowell, L. 1983: Towards an understanding of the gender division of urban space. *Environment and Planning D: Society and Space* 1, 59–72.
—— 1989: Gender divisions. In Hamnett, C., McDowell, L. and Sarre, P. (eds.), *The Changing Social Structure*. London: Sage Publications, 158–98.

—— 1993: Power and masculinity in city work places. Paper presented at the annual conference of the Institute of British Geographers.
McLuhan, M. 1964: *Understanding Media: The Extensions of Man*. London: Routledge and Kegan Paul.
Meinig, D. (ed.) 1979: *The Interpretation of Ordinary Landscapes*. New York: Oxford University Press.
Mellencamp, P. 1990: *Indiscretions: Avant-Garde Film, Video and Feminism*. Bloomington: Indiana University Press.
Mercer, K. and Julien, I. 1988: Race, sexual politics and black masculinity: a dossier. In Chapman, R. and Rutherford, J. (eds.), *Male Order: Unwrapping Masculinity*. London: Lawrence and Wishart, 97–164.
Merquior, J. 1989: *Foucault*. Fontana: London.
Merton, R. 1957: *Social Theory and Social Structure*. Glencoe, IL: Free Press.
Meyerhoff, B. 1975: Organisation and ecstacy: deliberate and accidental communitas among Huichol Indians and American youth. In Moore, S. and Meyerhoff, B. (eds.), *Symbol and Practice in Communal Ideology*. Ithaca and London: Cornell University Press.
Middleton, P. 1992: *The Inward Gaze: Masculinity and Subjectivity in Modern Culture*. London: Routledge.
Miller, H. 1956: *Quiet Days in Clichy*. London: Allison and Busby.
Mills, C. 1988: 'Life on the upslope': The postmodern landscapes of gentrification. *Environment and Planning D: Society and Space* 6, 169–89.
—— 1993: Myths and meanings of gentrification. In Duncan, J. and Ley, D. (eds.), *Place/Culture/Representation*. London: Routledge.
Minh-ha, T. T. 1989: *Woman, Native, Other: Writing, Postcoloniality and Feminism*. Bloomington: Indiana University Press.
Mitchell, D. 1992: Iconography and locational conflict from the underside: free speech, people's park and the politics of homelessness in Berkeley, California. *Political Geography* 11, 152–69.
Mitchell, J. 1974: *Psychoanalysis and Feminism*. Harmondsworth: Penguin.
Mitchell, J. and Rose, J. (eds.) 1982: *Feminine Sexuality: Jacques Lacan and the Ecole Freudienne*. London: Macmillan.
Monteith, M. 1991: Doris Lessing and the politics of violence. In Armitt, L. (ed.), *Where No Man Has Gone Before: Women and Science Fiction*. London: Routledge.
Moody, N. 1991: Maeve and Guinevere: women's fantasy writing in the science fiction marketplace. In Armitt, L. (ed.), *Where No Man Has Gone Before: Women and Science Fiction*. London: Routledge.
Moore, H. 1990: Visions of the good life: anthropology and the study of utopia. *Cambridge Anthropology* 14, 13–33.
Moos, A. 1989: The grassroots in action: gays and seniors capture the local state in West Hollywood, California. In Wolch, J. and Dear, M. (eds.), *The Power of Geography: How Territory Shapes Social Life*. Boston: Unwin Hyman.
More, T. 1516 and 1964: (Surtz, E., ed.), *Utopia*. New York: Yale University Press.
Morgan, D. 1987: *It Will Make a Man of You: Notes on National Service, Masculinity and Autobiography*. Manchester: University of Manchester.
Morley, D. 1991: Where the global meets the local: notes from the sitting room. *Screen* 32, 1–14.
Morley, D. and Robins, K. 1992: Techno-orientalism: foreigners, phobias and futures. *New Formations* 16, 136–56.
Morris, L. 1993: Is there a British underclass? *International Journal of Urban and Regional Research* 17, 404–12.
Mort, F. 1987: *Dangerous Sexualities: Medico-Moral Politics in England since 1830*. London: Routledge and Kegan Paul.

—— 1988: Boy's own? Masculinity, style and popular culture. In Chapman, R. and Rutherford, J. (eds.), *Male Order: Unwrapping Masculinity*. London: Lawrence and Wishart, 193–224.
Moser, B. 1986: Amazonian Indians (letter). *Anthropology Today* 2.1.
Mulvey, L. 1975: Visual pleasure and narrative cinema. *Screen* 16, 6–18.
—— 1989: *Visual and Other Pleasures*. Basingstoke: Macmillan.
Murray, C. 1990: *The Emerging British Underclass*. London: IEA Health and Welfare Unit.
Murphy, D. 1965: *Full Tilt: Durkirk to Delhi by Bicycle*. London: Arrow Books.
Myrdal, G. 1962: *The Challenge to Affluence*. New York: Pantheon.
Neale, S. 1983: Masculinity as spectacle: reflections on men and mainstream cinema. *Screen* 24, 2–17.
Nicol, T. 1991: Rock into urban geography. *The South Hampshire Geographer* 20, 17–24.
Nichols, G. 1989: Wherever I hang. In Nichols, G. (ed.), *Lazy Thoughts of a Lazy Woman and Other Poems*. London: Virago, 10.
Nietzsche, F. 1980: *On the Advantage and Disadvantage of History for Life*. Indianapolis: Hackett.
Nin, A. 1950: *The Four-Chambered Heart*. London: Virago.
—— 1974: *The Diaries of Anaïs Nin*. London: Harvest/HBJ.
—— 1986: *Henry and June*. Harmondsworth: Penguin.
Nixon, S. 1992: Have you got the look? Masculinities and shopping spectacle. In Shields, R. (ed.), *Lifestyle Shopping: The Subject of Consumption*. London: Routledge, 149–69.
Norris, C. 1990: *What's Wrong with Postmodernism*. London: Harvester Wheatsheaf.
Ogborn, M. 1993: Law and discipline in nineteenth century English state formation: the Contagious Diseases Acts of 1864, 1866 and 1869. *Journal of Historical Sociology* 6, 28–54.
O'Hanlon, R. 1984: Into the heart of Borneo. *Granta* 10, 59–82.
Okely, J. and Callaway, H. (eds.) 1992: *Anthropology and Autobiography*. London: Routledge.
Olsson, G. 1992: Lines of power. In Barnes, T. and Duncan, J. (eds.), *Writing Worlds: Discourse, Text and Metaphor in the Representation of Landscape*. London: Routledge.
Orbaum, J. 1992: *Sega Pro-Master, Vol. 1*. London: Simon and Schuster.
Orwell, G. 1949: *Nineteen Eighty-Four*. London: Secker and Warburg.
Partington, A. 1989: The designer housewife in the 1950s. In Attfield, J. and Kirkham, P. (eds.), *A View from the Interior: Feminism, Women and Design*. London: The Women's Press.
Peet, R. 1986: World capitalism and the destruction of regional cultures. In Johnston, R. and Taylor, P. (eds.), *A World in Crisis: Geographical Perspectives*. Oxford: Blackwell.
—— 1991: *Global Capitalism: Theories of Societal Development*. London: Routledge.
Pettigrew, J. 1975: *Rober Noblemen: A Study of the Political System of the Sikh Jats*. London: Routledge.
Philo, C. 1989: 'Enough to drive one mad': the organization of space in 19th-century lunatic asylums. In Wolch, J. and Dear, M. (eds.), *The Power of Geography: How Territory Shapes Social Life*. Boston: Unwin Hyman.
—— 1992: Foucault's geography. *Environment and Planning D: Society and Space* 10, 137–61.
Piercy, M. 1979: *Woman on the Edge of Time*. London: The Woman's Press.
Pile, S. 1993: Human agency and human geography revisited: a critique of 'new

models' of the self. *Transactions of the Institute of British Geographers* 18, 122–39.
Plant, S. 1992: *The Most Radical Gesture: The Situationist International in a Post-Modern Age*. London: Routledge.
Plester, P. 1993: Sticker: a Cornish community. Unpublished B.A. (Hons.) dissertation. Department of Geography, University of Portsmouth.
Poewe, K. 1988: *Childhood in Germany During World War II: The Story of a Little Girl*. Lewiston, Queenston and Lampeter: The Edward Mellen Press.
Poovey, M. 1990: Speaking of the Body: mid-Victorian constructions of female desire. In Jacobus, M., Keller, E. and Shuttleworth, S. (eds.), *Body/Politics: Women and Discourses of Science*, pp.29–46. London: Routledge.
Porteous, J. 1987: Inscape: landscapes of the mind in the Canadian and Mexican novels of Malcolm Lowry. *The Canadian Geographer* 30, 123–31.
Poster, M. 1984: *Foucault, Marxism and History*. Cambridge: Polity Press.
—— 1991: War and the mode of information. *Cultural Critique* (Fall), 217–22.
Pratt, M. 1992: *Imperial Eyes: Travel Writing and Transculturation*. London: Routledge.
Pulsipher, L. 1993: 'He won't let she stretch she foot': Gender relations in traditional West Indian houseyards. In Katz, C. and Monk, J. (eds.), *Full Circles: Geography of Women Over the Life Course*. London: Routledge.
Raban, J. 1974: *Soft City*. London: Hamilton.
—— 1990: *Hunting Mr Heartbreak*. London: Picador.
Rabinow, P. (ed.) 1984: *The Foucault Reader*. Harmondsworth: Penguin.
Rahimian, A., Wolch, J. and Koegel, P. 1992: A model of homeless migration: homeless men in Skid Row, Los Angeles. *Environment and Planning A*, 1317–36.
Relph, E. 1976: *Place and Placelessness*. London: Pion.
—— 1981: *Rational Landscapes and Humanistic Geography*. London: Croom Helm.
—— 1987: *The Modern Urban Landscape*. London: Croom Helm.
Richards, A. 1982: *Chisungu: A Girl's Initiation Ceremony among the Bemba of Zambia*. London: Tavistock.
Ricoeur, P. 1970: *Freud and Philosophy: An Essay on Interpretation*. Newhaven: Yale University Press.
—— 1978: *The Rule of Metaphor: Multidisciplinary Studies in the Creation of Meaning in Language*. London: Routledge.
—— 1986: *Lectures on Ideology and Utopia*. New York: Columbia University Press.
Roper, M. and Tosh, J. (eds.) 1991: *Manful Assertions: Masculinities in Britain Since 1800*. London: Routledge.
Rose, G. 1993a: *Feminism and Geography: The Limits of Geographical Knowledge*. Cambridge: Polity Press.
—— 1993b: Some notes towards thinking about the spaces of the future. In Bird, J., Curtis, B., Putnam, T., Robertson, G. and Tickner, L. (eds.), *Mapping the Futures: Local Cultures, Global Change*. London: Routledge.
Rose, N. 1990: *Governing the Soul: The Shaping of the Private Self*. London: Routledge.
—— 1992: Governing the enterprising self. In Heelas, P. and Morris, P. (eds.), *The Values of the Enterprise Culture: The Moral Debate*. London: Routledge.
Rowbotham, S. 1985: Revolt in Roundhay. In Heron, L. (ed.), *Truth, Dare or Promise: Girls Growing Up in the Fifties*. London: Virago.
Rowe, S. and Wolch, J. 1990: Social networks and in time and space: homeless women in Skid Row, Los Angeles. *Annals of the Association of American Geographers* 80, 184–204.

Rushdie, S. 1990: *In Good Faith*. London: Granta Books.
—— 1991: *Imaginary Homelands: Essays and Criticism 1981–1991*. London: Granta Books.
Russ, J. 1975: *The Female Man*. Boston: Beacon Press.
Rutherford, J. 1988: Who's that man? In Chapman, R. and Rutherford, J. (eds.), *Male Order: Unwrapping Masculinity*. London: Lawrence and Wishart, 21–67.
—— 1992: *Men's Silences: Predicaments in Masculinity*. London: Routledge.
Sacquin, M. and Ladurie, E. 1993: *Le Printemps des genies: les enfants prodiges*. Paris: Bibliotheque Nationale/Robert Laffont.
Said, E. 1978: *Orientalism*. New York: Columbia University Press.
—— 1993: *Culture and Imperialism*. London: Chatto and Windus.
Sand, G. 1929: *The Intimate Journal of George Sand*. Chicago: Academy Press.
Sauer, C. 1963: The morphology of landscape. In Leighly, J. (ed.), *Land and Life: A Selection from the Writing of Carl Otwin Sauer*. Berkeley: University of California Press.
Saunders, P. 1986: *Social Theory and the Urban Question*, 2nd edition. London: Hutchinson Education.
—— 1989: The meaning of 'home' in contemporary English culture. *Housing Studies* 4, 177–92.
Schildkrout, E. 1978: Roles of children in urban Kano. In La Fontaine, J. (ed.), *Age and Sex as Principles of Social Difference*. London, New York and San Francisco: Academic Press.
Schuster, I. 1979: *The New Women of Lusaka*. Palo Alto: Mayfield.
Seamon, D. 1985: Reconciling old and new worlds: the dwelling–journey relationship as portrayed in Vilhelm Moberg's 'emigrant' novels. In Seamon, D. and Mugerauer, R. (eds.), *Dwelling, Place and Environment: Towards a Phenomenology of Person and World*. Dordrecht: Martinus Nijhoff.
Seamon, D. and Mugerauer, R. (eds.) 1985: *Dwelling, Place and Environment: Towards a Phenomenology of Person and World*. Dordrecht: Martinus Nijhoff.
Segalen, M. 1991: *Fifteen Generations of Bretons: Kinship and Society in Lower Brittany 1720–1980*. Cambridge: Cambridge University Press.
Sen, A. 1987: *The Standard of Living*. Cambridge: Cambridge University Press.
Shields, R. 1989: Social spatialization and the built environment: the West Edmonton Mall. *Environment and Planning D: Society and Space* 7, 147–64.
—— (ed.) 1991a: *Lifestyle Shopping: The Subject of Consumption*. London: Routledge.
—— 1991b: *Places on the Margin: Alternative Geographies of Modernity*. London: Routledge.
Showalter, E. 1992: The way we write now: syphilis and AIDS. In Showalter, E. (ed.), *Sexual Anarchy: Gender and Culture at the Fin de Siècle*. London: Virago.
Shumway, D. 1989: *Michel Foucault*. Charlottesville: University of Virginia Press.
Shurmer, P. 1972: The gift game. *New Society* 19/12.
Shurmer-Smith, P. 1984: The Sikh Identity. *Geographical Magazine* September, 442–3.
—— 1991: Ideal homes: two islands in the construction of family identity. In Philo, C. (compiler). *New Words, New Worlds: Reconceptualising Social and Cultural Geography*. Lampeter: IBG Social and Cultural Study Group.
—— 1993: Romances of remoteness: the place of the Highlands and Islands in imagination and practice. In Mukherjee, A. and Agnihotri, V. (eds.), *Environment and Development: Views from the East and the West*. New Delhi: Concept.
—— 1994: Cixous' spaces. *Ecumene* 1.4.
Simmel, G. 1971: Metropolis and mental life. In Levine, D. (ed.), *Individuality and Social Forms*. Chicago: Chicago University Press.

—— 1990: *The Philosophy of Money*. London: Routledge.
Simpson, D. 1990: Destiny made manifest: the styles of Whitman's poetry. In Bhabha, H. (ed.), *Nation and Narration*. London: Routledge, 177–96.
Smith, D. J. (ed.) 1992: *Understanding the Underclass*. London: Policy Studies Institute.
Smith, N. 1992: Blind man's buff, or Hamnett's philosophical individualism in search of gentrification. *Transactions of the Institute of British Geographers* 17, 110–15.
—— 1993: Homeless/global: scaling places. In Bird, J., Curtis, B., Putnam, T., Robertson, G. and Tickner, L. (eds.), *Mapping the Futures: Local Cultures, Global Change*. London: Routledge.
Soja, E. 1989: *Postmodern Geographies*. London: Verso.
Somerville, P. 1992: Homelessness and the meaning of home: rooflessness or rootlessness. *International Journal for Urban and Regional Research* 16, 529–39.
Spivak, G. 1987: *In Other Worlds: Essays in Cultural Politics*. New York: Methuen.
—— 1990: An interview. *Radical Philosophy* 54, 31–4.
Spivak, G. and Haraym (eds.) 1990: *The Post-Colonial Critic*. London: Routledge.
Srinivas, M. 1977: *The Remembered Village*. Delhi: Oxford University Press.
Stallybrass, P. and White, A. 1986: *The Politics and Poetics of Transgression*. London: Methuen.
Standing, H. 1991: *Dependence and Autonomy: Women's Employment and the Family in Calcutta*. London: Routledge.
Stanley Robinson, K. 1992: *Red Mars*. London: HarperCollins.
—— 1993: *Green Mars*. London: HarperCollins.
Steinfield, E. 1981: The place of old age: the meaning of housing for old people. In Duncan, J. (ed.), *Housing and Identity: Cross Cultural Perspectives*. London: Croom Helm.
Stock, B. 1993: Reading, community and sense of place. In Duncan, J. and Ley, D. (eds.), *Place/Culture/Representation*. London: Routledge.
Stocking, G. Jr. 1987: *Victorian Anthropology*. New York: The Free Press.
Stump, R. 1991: Spatial implications of religious broadcasting: stability and change in patterns of belief. In Brunn, S. and Leinbach, T. (eds.), *Collapsing Space and Time: Geographic Aspects of Communication and Information*. London: HarperCollins Academic.
Süskind, P. 1988: *The Pigeon*. Harmondsworth: Penguin.
Telotte, J. 1990: The doubles of fantasy and the space of desire. In A. Kuhn (ed.), *Alien Zone*. London: Verso, 152–9.
Theroux, P. 1975: *Great Railway Bazaar*. Harmondsworth: Penguin.
Thompson, M. 1979: *Rubbish Theory: The Creation and Destruction of Value*. Oxford: Oxford University Press.
Thrift, N. 1989a: Images of social change. In Hamnett, C., McDowell, L. and Sarre, P. (eds.), *The Changing Social Structure*. London: Sage Publications, 12–42.
—— 1989b: The geography of international economic disorder. In Johnston, R. and Taylor, P. (eds.), *A World in Crisis? Geographical Perspectives*. Oxford: Blackwell, 16–78.
Thrift, N. 1993: For a new regional geography, 3. *Progress in Human Geography* 17, 92–100.
Tivers, J. 1978: How the other half lives: the geographical study of women. *Area* 10, 302–6.
—— 1985: *Women Attached: The Daily Lives of Women with Young Children*. London: Croom Helm.

Tönnies, F. 1955: *Community and Association.* London: Routledge and Kegan Paul.

Trevor-Roper, H. 1983: The invention of tradition: the Highland tradition of Scotland. In Hobsbawm, E. and Ranger, T. (eds.). *The Invention of Tradition.* Cambridge: Cambridge University Press.

Trigger, B. 1988: Reply to Julia Harrison's '"The spirit sings" and the future of anthropology'. *Anthropology Today* 4.6.

Trollope, J. 1991: *The Rector's Wife.* London: Black Swan.

Tuan, Y.-F. 1974: *Topophilia: A Study of Environmental Perception.* London: Prentice Hall.

—— 1993: *Passing, Strange and Wonderful: Aesthetics, Nature and Culture.* Washington DC: Island Press.

Turnbull, C. 1985: *The Human Cycle.* London: Triad Paladin.

Turner, L. and Ash, J. 1975: *The Golden Hordes.* London: Constable.

Turner, V. 1969: *The Ritual Process: Structure and Anti-Structure.* London: Routledge.

—— 1974: *Dramas, Fields and Metaphors: Symbolic Action in Human Society.* Ithaca: Cornell University Press.

Turner, V. and Turner, E. 1978: *Image and Pilgrimage in Christian Culture.* Oxford: Blackwell.

Urry, J. 1988: Cultural change and contemporary holiday-making. *Theory, Culture and Society* 5, 35–55.

—— 1990: *The Tourist Gaze: Leisure and Travel in Contemporary Societies.* London: Sage Publications.

Valentine, C. 1968: *Culture and Poverty: Critique and Counter-Proposals.* Chicago: University of Chicago Press.

Valentine, G. 1992: Images of danger: women's sources of information about the spatial distribution of male violence. *Area* 24, 22–9.

—— 1993: Negotiating and managing multiple sexual identities: lesbian time-space strategies. *Transactions of the Institute of British Geographers* 18, 237–48.

Veness, A. 1992: Home and homelessness in the United States: changing ideals and realities. *Environment and Planning D: Society and Space* 10, 445–68.

—— 1993: Neither homed, nor homeless: contested definitions and the personal worlds of the poor. *Political Geography* 12, 319–40.

Virilio, P. 1984: *L'espace critique.* Paris: Christian Bourgeois.

—— 1986: The overexposed city. *Zone* 1/2, 14–39.

—— 1989: *War and Cinema: The Logistics of Perception.* London: Verso.

—— 1991: *The Lost Dimension.* New York: Semiotext(e).

Wacquant, L. 1993: Urban outcasts: stigma and division in the black American ghetto and the French urban periphery. *International Journal of Urban and Regional Research* 17, 384–404.

Wacquant, L. and Wilson, W. 1989: The cost of racial and class exclusion in the inner city. In Wilson, W. (ed.), *The Ghetto Underclass.* London: Sage Publications.

Wagner, R. 1986: *Symbols that Stand for Themselves.* London: University of Chicago Press.

Walker, M. 1993: Mickey Mouse culture. *The Guardian*, 27 December, 18.

Walkerdine, V. 1985: Dreams from an ordinary childhood. In Heron, L. (ed.), *Truth, Dare or Promise: Girls Growing Up in the Fifties.* London: Virago.

Wallis, R. and Baran, S. 1990: *The Known World of Broadcast News: International News and the Electronic Media.* London: Routledge.

Warner, M. 1985: *Monuments and Maidens: The Allegory of the Female Form.* London: Picador.

Westwood, S. 1990: Racism, black masculinity and the politics of space. In Hearn, J. and Morgan, D. (eds.), *Men, Masculinities and Social Theory*. London: Unwin Hyman.
White, P. 1992: Female spectator, lesbian specter: The Haunting. In Colomina, B. (ed.), *Sexuality and Space*. New York: Princeton Architectural Press.
Whitman, W. 1912: O Captain! My Captain! In Whitman, W., *Leaves of Grass*. London: J. M. Dent, 282.
Wilde, O. 1973: (Pearson, H., ed.) *De Profundis and Other Essays*. Harmondsworth: Penguin Books.
Willetts, D. 1992: Theories and explanations of the underclass. In Smith, D. J. (ed.), *Understanding the Underclass*. London: Policy Studies Institute.
Williams, R. 1975: *The Country and the City*. London: Paladin.
—— 1976: *Keywords: A Vocabulary of Culture and Society*. London: Fontana.
—— 1983: *Towards 2000*. London: Chatto and Windus.
Wilson, E. 1991: *The sphinx in the city: urban life, the control of disorder, and women*. London: Virago.
Wilson, W. 1987: *The Truly Disadvantaged, the Inner City, the Underclass and Public Policy*. Chicago: University of Chicago Press.
—— (ed.) 1989: *The Ghetto Underclass*. London: Sage Publications.
Wirth, L. 1964: *On Cities and Social Life*. Chicago: University of Chicago Press.
Wolf, N. 1990: *The Beauty Myth: How Images of Beauty are used against Women*. London: Vintage.
Wolfe, T. 1940 and 1962: *You Can't Go Home Again*. Harmondsworth: Penguin.
Wolch, J., Koegel, P. and Rahimian, A. 1993: Daily and periodic mobility patterns of the urban homeless. *Professional Geographer* 45, 159–69.
Woolf, V. 1945: *A Room of One's Own*. Harmondsworth: Penguin.
Worseley, P. 1957: *The Trumpet Shall Sound*. London: Paladin.
Wright, P. 1985: *On Living in an Old Country*. London: Verso.
Wright, S. 1993: Blaming the victim, blaming society or blaming the discipline: fixing responsibility for poverty and homelessness. *The Sociological Quarterly* 34, 1–16.
Young, M. 1991: *An Inside Job: Policing and Police Culture in Britain*. Oxford: Oxford University Press.
Zamyatin, Y. 1924: *We*. New York: Dutton.
Zonabend, F. 1984: (tr. Forster, A.) *The Enduring Memory: Time and History in a French Village*. Manchester: Manchester University Press.
—— 1993: *The Nuclear Peninsular*. Cambridge: Cambridge University Press.
Zukin, S. 1992: Postmodern urban landscapes: mapping culture and power. In Friedman, J. and Lash, S. (eds.), *Modernity and Identity*. Oxford: Blackwell, 221–47.

Index